FLUID DYNAMIC APPLICATIONS
OF THE DISCRETE BOLTZMANN EQUATION

Series on Advances in Mathematics for Applied Sciences

*To view the complete list of the published volumes in the series, please visit:
https://www.worldscibooks.com/series/samas_series.shtml

Series on Advances
in Mathematics for
Applied Sciences
Vol. 3

FLUID DYNAMIC APPLICATIONS OF THE DISCRETE BOLTZMANN EQUATION

Roberto Monaco & Luigi Preziosi
Dipartimento di Matematica
Politecnico di Torino

World Scientific
Singapore • New Jersey • London • Hong Kong

Published by

World Scientific Publishing Co. Pte. Ltd.
5 Toh Tuck Link, Singapore 596224
USA office: 27 Warren Street, Suite 401-402, Hackensack, NJ 07601
UK office: 57 Shelton Street, Covent Garden, London WC2H 9HE

British Library Cataloguing-in-Publication Data
A catalogue record for this book is available from the British Library.

Series on Advances in Mathematics for Applied Sciences — Vol. 3
FLUID DYNAMIC APPLICATIONS OF THE DISCRETE
BOLTZMANN EQUATION

Copyright © 1991 by World Scientific Publishing Co. Pte. Ltd.

ISBN-13   978-981-02-0466-2
ISBN-10   981-02-0466-3

Acknowledgements

The authors acknowledge the partial support of both the Italian Ministry for University, Scientific and Technological Research and the National Research Council (Gruppo Nazionale Fisica Matematica).

PREFACE

The discrete Boltzmann equation is a mathematical model, or a class of models, of the nonlinear kinetic theory for a gas of point mass particles which can only attain to a finite number of velocities.

The idea of discretizing the velocity space certainly belongs to Maxwell, however the first realistic step is due to Broadwell [1,2], who proposed a simple model of a gas with six velocities only, all with the same modulus, directed along the positive and negative directions of a Carthesian orthogonal frame.

Broadwell's model was successfully applied to the analysis of shock wave propagation and steady Couette flow. He was able to obtain analytic solutions and consequently a nice and accurate description of the flow pattern. These first successful approaches encouraged the development of a discrete kinetic theory, modelling, thermodynamics and application. This research result is the one contained in the fundamental Lecture Notes by Gatignol [3], which still remains the starting point of all research activity in the field.

The interest in a discrete kinetic theory is also documented in the volume by Sultanghazin [4] and in several fluid-dynamic applications which will be reviewed in this book. In particular the original theory referred to a simple monoatomic gas of equal particles undergoing binary elastic collisions has been developed in order to include several interesting physical descriptions. A discrete kinetic theory for gas mixtures is developed in [5] starting from the work of Bellomo and de Socio [6]. A systematic analysis of triple collision and dense gas effects is developed in [7,8]. A discrete kinetic theory for chemically reacting gases is developed and applied to fluid-dynamics in [9,10]. Moreover discrete kinetic theory can be regarded as the natural framework of lattice gasdynamics [11].

Therefore the original motivations of a discrete kinetic theory, mainly consisting in the modelling of a gas particles sufficiently simple to provide immediate analytic descriptions of flow patterns, have been modified addressing the research activity towards a discrete kinetic theory related to general physical situations with the aim of providing a tool useful for fluid-dynamics.

All these motivations have suggested to produce this volume which is essentially oriented to fluid-dynamic applications in interesting and general physical situations. Modelling and thermodynamic aspects are dealt with in the first two chapters. Mathematical theory and methods are reviewed in the third chapter. The analysis of shock wave phenomena is developed in the fourth chapter. Chapter 5 deals with the analysis of fluid-dynamic problems in bounded domains and of flows over convex bodies. The last chapter deals with the discrete kinetic theory and its application for chemically reacting gases. Each chapter indicates possible developments and open research problems.

We need mentioning, before concluding this preface, that mathematicians have been strongly attracted by this class of models with the aim of providing useful theorems in the analysis of the initial and initial-boundary value problem. The pioneers papers in this line are the existence results for the initial value problem for the Broadwell model in one space dimension by Nishida and Mimura [12], developed to more general models by Cabannes [13], to three space dimensions by Kawashima [14] and to an existence theory for large initial data by Tartar [15]. All papers devoted to mathematical analysis are reviewed in the survey papers by Platkowski and Illner [16] and Bellomo and Gustafsson [17]. However, it seems, at present, that only in one space dimensions it is possible to obtain for the discrete Boltzmann equation stronger mathematical results than the corresponding ones referred to the continuous Boltzmann equation [18,19].

Considering that this volume essentially deals with fluid-dynamics, the existence theory is not dealt with in the volume itself. However some results on the qualitative analysis, which can be regarded useful for the applications, will be reviewed and discussed in Chapter 3.

This book is essentially addressed to applied mathematicians who work on non-linear problems in kinetic theory of gases and related areas. Moreover, the book could be utilized by postgraduate students who want to enter this research field and by engineers who work in gasdynamics and aerospace sciences.

Although this book tries to be self-contained and to refer as small as possible to the existing literature, in order to understand the parallelism with the continuous Boltzmann equation, the reader should be acquainted with the preliminary notions of kinetic theory, say Chapters 1 and 2 of [20].

References

[1] I.E. Broadwell, "Shock structure in a simple discrete velocity gas", *Phys. Fluids*, **7**, 1964, p.1243.

[2] I.E. Broadwell, "Study of rarefied shear flow by the discrete velocity method", *J. Fluid Mech.*, **19**, 1964, p.401.

[3] R. Gatignol, **Théorie Cinétique des Gaz a Répartition Discrète de Vitesses**, *Lecture Notes in Phys.* No.**36**, Springer-Verlag, Berlin, New York, 1975.

[4] U.M. Sultanghazin, **Discrete Nonlinear Models of the Boltzmann Equation**, Nauka Publ., Moskow, 1987.

[5] E. Longo and R. Monaco, "On the thermodynamics of the discrete models of the Boltzmann equation for gas mixtures", *Transp. Theory Statist. Phys.*, **17**, 1988, p.423.

[6] N. Bellomo and L. M. de Socio, "The discrete Boltzmann equation for gas mixtures. A regular space model and shock wave problem",

Mech. Res. Comm., **10**, 1983, p.233.

[7] R. Gatignol and F. Coulouvrat, "Description hydrodinamique d'un gaz en théorie cinétique discrète: le modèl général", *Comp. Rend. Acad. Sci. Paris, I*, **306**, 1988, p.169.

[8] N. Bellomo and S. Kawashima, "The discrete Boltzmann equation with multiple collisions: Global existence and stability for the initial value problem", *J. Math. Phys.*, **31**, 1990, p.245.

[9] E. Gabetta and R. Monaco, "The discrete Boltzmann equation for gases with bi-molecular chemical reactions", in **Discrete Models of Fluid Dynamics**, *Advances in Mathematics for Applied Sciences* vol.**2**, Ed. A. Alves, World Scientific, London, Singapore, p.22.

[10] R. Monaco and M. Pandolfi Bianchi, "Shock wave onset with chemical dissociation by the discrete Boltzmann equation", in **Rarefied Gas Dynamics**, Ed. A. Beylich, VCH-Verlag, Weinheim, New York, 1991, p.862.

[11] **Proc. of Discrete Kinetic Theory, Lattice Gas Dynamics and Foundations of Hydrodynamics**, Ed. R. Monaco, World Scientific, London, Singapore, 1989.

[12] T. Nishida and M. Mimura, "On the Broadwell's model for a simple discrete velocity gas", *Proc. Japan Acad.*, **50**, 1974, p.812.

[13] H. Cabannes, "Global solution of the discrete Boltzmann equation", in **Mathematical Problems in the Kinetic Theory of Gases**, Eds. H. Neunzert and D. Pack, Lang Publ. Co., Frankfurt, 1980, p.25.

[14] S. Kawashima, "The asymptotic equivalence of the Broadwell model

equation and its Navier-Stokes equation", *Japan J. Math.*, **7**, 1981, p.1.

[15] L. Tartar, "Existence globale pur un système hyperbolique semilinéaire de la théorie cinétique des gaz", *Séminaire Goulaouic–Schwartz*, **1**, 1975.

[16] T. Platkowski and R. Illner, "Discrete velocity models of the Boltzmann equation: A survey on the mathematical aspects of the theory", *SIAM Review*, **30**, 1988, p.213.

[17] N. Bellomo and T. Gustafsson, "The discrete Boltzmann equation: A review of the mathematical aspects of the initial and initial-boundary value problem", *Review Math. Phys.*, to appear in 1991.

[18] N. Bellomo, A. Palczewski and G. Toscani, **Mathematical Topics in Nonlinear Kinetic Theory**, World Scientific, London, Singapore, 1988.

[19] R. DiPerna and P.L. Lions, "On the Cauchy problem for the Boltzmann equation: Global existence and weak stability", *Annals of Math.*, **130**, 1989, p.321.

[20] M.N. Kogan, **Rarefied Gas Dynamics**, Plenum Press, New York, 1969.

CONTENTS

Chapter 2: SOME DISCRETE VELOCITY MODELS

Chapter 3: ON THE DISCRETE BOLTZMANN EQUATION IN UNBOUNDED DOMAINS

CHAPTER 1

THE DISCRETE BOLTZMANN EQUATION

MODELLING AND THERMODYNAMICS

The discrete Boltzmann equation (DBE) is a nonlinear mathematical model of the kinetic theory of gases [1], that describes the time-space
evolution of a gas of particles which are allowed to move with a finite
number of velocities only. The model is an evolution equation for the
number densities linked to the selected velocities.

Simple models, with a limited number of velocities, have been proposed by Broadwell [2,3] and applied to the analysis of classical fluid-
dynamic problems: Couette flows and shock wave phenomena.

However the discrete kinetic theory was developed in a systematic
and rigorous fashion only in the Lecture Notes [1]. Gatignol's volume
[1] develops a detailed analysis of the relevant aspects of the discrete
kinetic theory: modelling, analysis of thermodynamic equilibrium and

application to fluid-dynamic problems.

The Lecture Notes [1] mainly refer to a simple monoatomic gas and to the related thermodynamic aspects. After the fundamental contribution of [1], several progresses have been made. These recent results motivate this present volume, mainly oriented towards the analysis of fluid-dynamic problems. In fact, the discrete Boltzmann equation has been developed, as we shall see, for gas mixtures, for gases with multiple collisions and for chemically reacting gases. Therefore the model can be regarded as an interesting one for fluid-dynamic applications.

Even if this present volume deals with fluid-dynamic problems, we mention that the discrete Boltzmann equation has stirred up the interest of applied mathematicians (as documented in the review papers by Platkowski and Illner [4] and Bellomo and Gustafsson [5]) who tackled the problem of providing an existence theory to classical initial and initial-boundary value problems.

The knowledge of the content of [1,4,5] will certainly be useful to follow this volume, which, however, is organized in a self-contained form with somewhat limited reference to the existing literature.

This first Chapter is organized in five additional (to this introduction) sections. Section 1 provides the formal structure of the discrete Boltzmann equation. Section 2 deals with some methodological aspects of the mathematical modelling of the equation. Section 3 indicates how to compute the macroscopic observables from the knowledge of the microscopic behaviour of the gas, namely the densities N_i. Section 4 deals with the analysis of the equilibrium Maxwellian state and finally, the last section with the presentation of the plan of the book.

1.1   The Discrete Boltzmann Equation

As already mentioned, the discrete Boltzmann equation is a mathematical model of the nonlinear kinetic theory of gases that describes the evolution of a gas of particles which are allowed to move only with a finite number of velocities $\mathbf{v}_i$, $i = 1, \ldots, n$.

The particles move in the whole space and collide by simple elastic collisions locally in space. The mathematical model is an evolution equation for the densities N_i linked to the velocities $\mathbf{v}_i$.

Let t and $\mathbf{x} \in \mathbf{R}^d$, $d = 1, 2, 3$ be respectively the time and the space variables. Then if $N_i = N_i(t, \mathbf{x})$ is the number density referred to the velocity $\mathbf{v}_i \in \mathbf{R}^d$, then the formal expression of the discrete Boltzmann equation with both binary and triple collisions is

$$N_i = N_i(t, \mathbf{x}) : (t, \mathbf{x}) \in [0, T] \times \mathbf{R}^d \longmapsto \mathbf{R}_+ \quad , \quad i = 1, \ldots, n$$

$$(\frac{\partial}{\partial t} + \mathbf{v}_i \cdot \nabla_{\mathbf{x}}) N_i = J_i[\mathbf{N}] = J_i^{(2)}[\mathbf{N}] + J_i^{(3)}[\mathbf{N}] \ , \qquad (1.1)$$

where $J_i^{(2)}$ and $J_i^{(3)}$ are, respectively, the nonlinear binary and triple collision terms which, with the microreversibility property which will be given afterwards, re-write

$$J_i^{(2)}[\mathbf{N}] = \frac{1}{2} \sum_{jhk=1}^{n} A_{ij}^{hk}(N_h N_k - N_i N_j) \qquad (1.2)$$

$$J_i^{(3)}[\mathbf{N}] = \frac{1}{3!} \sum_{j\ell ghk=1}^{n} A_{ij\ell}^{ghk}(N_g N_h N_k - N_i N_j N_\ell) \ . \qquad (1.3)$$

The terms A are the so-called *transition rates* referred to the collisions

$$(\mathbf{v}_i, \mathbf{v}_j) \longleftrightarrow (\mathbf{v}_h, \mathbf{v}_k) \ , \qquad i, j, h, k = 1, \ldots, n \qquad (1.4)$$

in the case of binary collisions and

$$(\mathbf{v}_i, \mathbf{v}_j, \mathbf{v}_\ell) \longleftrightarrow (\mathbf{v}_g, \mathbf{v}_h, \mathbf{v}_k) \ , \qquad i, j, \ell, g, h, k = 1, \ldots, n \qquad (1.5)$$

in the case of triple collisions.

The transition rates are positive constants which, according to the indistinguishability property of the gas particles and to the reversibility of the collisions, satisfy the following relations

$$A_{ij}^{hk} = A_{ji}^{hk} = A_{ij}^{kh} = A_{ji}^{kh} \ ; \quad A_{ij}^{hk} = A_{hk}^{ij} \ . \qquad (1.6)$$

$$A_{ij\ell}^{ghk} = A_{i\ell j}^{ghk} = \cdots = A_{\ell ji}^{ghk} = \cdots = A_{ij\ell}^{khg} = \cdots = A_{\ell ji}^{khg} \ ; \quad A_{ij\ell}^{ghk} = A_{ghk}^{ij\ell} \ .$$
$$(1.7)$$

A detailed computation of the transition rates can be performed specializing the velocity discretization and analysing the related collision mechanics. This matter is dealt with in details in the next section.

1.2   On the Mathematical Modelling

Various aspects of the mathematical modelling of a simple monoatomic gas are dealt with in the Lecture Notes by Gatignol [1], modelling of gas mixtures are dealt with in [6,7] and modelling in the presence of triple collisions in [8–10].

The preceding section has already provided some information on the structure of the discrete Boltzmann equation and on some aspects of the mathematical modelling. This section deals with a unified presentation of the logic line which may be followed in the mathematical modelling. The presentation will refer to Section 1.1 and is organized in two paragraphs: first we deal with a simple one-component gas with both binary and triple collision terms and then with gas mixtures. The reader will find in Chapter 2 the detailed structure of several models which will be applied to specific fluid-dynamic problems in the chapters which follow.

The analysis of this chapter refers to a neutral gas. The mathematical modelling of the discrete Boltzmann equation with chemical reactions requires consistent modifications to the methodology proposed in this section and will be dealt with in Chapter 6.

1.2.1 *The Discrete Boltzmann Equation for a One-Component Gas*

Consider a simple monoatomic gas of particles of mass m and cross sectional area S. The first step of the modelling procedure consists in discretizing the velocity directions in a finite number of unit vectors

$$\mathbf{e}_i \in \mathbf{R}^d \ , \quad i = 1, \ldots, n \ . \tag{1.8}$$

Then one or more moduli are associated to each direction. The ratio of the moduli has, however, to be properly chosen, so that collisions between particles with different velocity moduli are possible. For instance, if only two velocity moduli are allowed, then one can introduce the following velocity discretization

$$
\mathbf{v}_i = \begin{cases} c\mathbf{e}_i & \text{if } i = 1,\ldots,n^* < n \\ \alpha c\mathbf{e}_i & \text{if } i = n^* + 1,\ldots,n \,, \end{cases} \tag{1.9}
$$

with $\alpha > 1$. The condition stated above requires that elastic, binary collisions of the type (1.4), such that momentum and energy are preserved

$$
\mathbf{v}_i + \mathbf{v}_j = \mathbf{v}_h + \mathbf{v}_k
$$
$$
|\mathbf{v}_i|^2 + |\mathbf{v}_j|^2 = |\mathbf{v}_h|^2 + |\mathbf{v}_k|^2 \,, \tag{1.10}
$$

are possible not only for $1 \leq i,j \leq n^*$ (or $n^* + 1 \leq i,j \leq n$), but also for $1 \leq i \leq n^*$ and $n^* + 1 \leq j \leq n$ (and vice versa).

The number of outputs $(\mathbf{v}_h, \mathbf{v}_k)$ corresponding to a given input $(\mathbf{v}_i, \mathbf{v}_j)$, consistent with the conservation equations (1.10), is denoted by q.

Once the discretization is stated and all the binary collisions satisfying (1.10) have been computed, the expression (1.2) of the collision operator is simply obtained by joining the transition rates A_{ij}^{hk} to the corresponding transition probability densities a_{ij}^{hk} through the relation

$$
A_{ij}^{hk} = S|\mathbf{v}_i - \mathbf{v}_j|a_{ij}^{hk} \,, \tag{1.11}
$$

where , of course,

$$a_{ij}^{hk} \geq 0 ,$$

$$\sum_{h,k=1}^{n} a_{ij}^{hk} = 1 , \quad \forall i, j = 1, \ldots, n .$$

If all q outputs referred to the input $(\mathbf{v}_i, \mathbf{v}_j)$ are assumed to be equally probable, then

$$a_{ij}^{hk} = \frac{1}{q} , \tag{1.12}$$

for all h and k such that the collision (1.4) satisfies (1.10), otherwise $a_{ij}^{hk} = 0$.

The term $S|\mathbf{v}_i - \mathbf{v}_j| dt$ is the volume spanned by the particle with velocity $\mathbf{v}_i$ in the relative motion with respect to the particle with velocity $\mathbf{v}_j$ in the time interval dt. Therefore

$$S|\mathbf{v}_i - \mathbf{v}_j| N_j$$

is the number of j-particles involved by the collision in the unit time.

The binary term in the collision operator is then the one defined in (1.2), where the transition rates are computed as it has been indicated above. In particular the positive contribution is defined *gain* term, whereas the negative one is defined *loss* term.

Note that the transition probability densities need to be characterized by the same reversibility and permutability properties already indicated for the rates A_{ij}^{hk} in (1.6). These properties always have to be verified in the analysis of the collision dynamics.

Collisions which satisfy the conservation and reversibility conditions which have been stated above are defined *admissible collisions*. Among these, the collisions with exchange of velocities

$$(\mathbf{v}_i, \mathbf{v}_j) \longleftrightarrow (\mathbf{v}_j, \mathbf{v}_i) \qquad i, j = 1, \ldots, n$$

are defined *trivial collisions* as the contribution to the collision operator is equal to zero, i.e. the gain and the loss terms in (1.2) are equal.

The discretization of velocity space indicated in (1.9) is used in several models known in the literature, e.g. Cabannes' model [11] with 14 velocities, 6 joining the center of a cube with the centers of its faces and 8 joining such a center with the vertices of the cube itself, or Gatignol's model [1] with $2 \times 2n$ velocities in the plane, all placed at the same angular distance, with $2n$ directions and two velocity moduli.

As already stated at the beginning of this section, the ratio of the moduli has to be properly chosen in order to allow collisions between particles with different moduli. Hence, in Cabannes' 14–velocity model $\alpha = \sqrt{3}$, while in Gatignol's $2 \times 2n$ model the ratio of the two moduli has to be chosen equal to

$$\alpha = \frac{1}{\sin \dfrac{\kappa \pi}{2n}} , \tag{1.13}$$

with $\kappa = 1$ if n is odd and $\kappa = 2$ if n is even.

The analysis of fluid-dynamic problems developed throughout this book will use models based upon the discretization (1.9), however it has to be mentioned that alternative types of discretization, with a large number of velocities, can be obtained as shown by Shizuta and Kawashima [12,13].

Referring now to the modelling of the triple collision terms, namely of the transition rates $A_{ij\ell}^{ghk}$ to be inserted into the expression of the collision operator $J_i^{(3)}[\mathbf{N}]$ in (1.3), one has to acknowledge that very little is known in the literature. As a matter of fact, despite several formal analysis of the equation, specific models have been proposed only in the case of *symmetric-type* collisions. The term *symmetric* is used, in particular, to denote collisions such that

$$\mathbf{v}_i + \mathbf{v}_j + \mathbf{v}_\ell = \mathbf{v}_g + \mathbf{v}_h + \mathbf{v}_k$$
$$|\mathbf{v}_i|^2 + |\mathbf{v}_j|^2 + |\mathbf{v}_\ell|^2 = |\mathbf{v}_g|^2 + |\mathbf{v}_h|^2 + |\mathbf{v}_k|^2 \ , \tag{1.14}$$

and the velocities are, in a plane, at the same angular distance.

In this physical situation, one can assume that a triple collision occurs if a third particle enters the *action volume* of two colliding ones, where the action volume is the one defined by the binary collision. Moreover, considering that all types of collisions

$$(\mathbf{v}_i, \mathbf{v}_j) \longleftrightarrow \mathbf{v}_\ell$$
$$(\mathbf{v}_i, \mathbf{v}_\ell) \longleftrightarrow \mathbf{v}_j$$
$$(\mathbf{v}_j, \mathbf{v}_\ell) \longleftrightarrow \mathbf{v}_i \ ,$$

are equally probable, the relation between the transition rates and probability densities is

$$A_{ij\ell}^{ghk} \propto S^{5/2} a_{ij\ell}^{ghk} \left(|\mathbf{v}_i - \mathbf{v}_j| + |\mathbf{v}_i - \mathbf{v}_\ell| + |\mathbf{v}_j - \mathbf{v}_\ell| \right) \ , \tag{1.15}$$

where $a_{ij\ell}^{ghk}$ is computed in the same way of the binary collisions and the constant of proportionality depends, as we shall see in Chapter 2, on the geometry of the model. The collision operator $J_i^{(3)}[\mathbf{N}]$ is then obtained inserting the expression of $A_{ij\ell}^{ghk}$ into Eq.(1.3)..

Examples of models which use formula (1.15) are given in Chapter 2. However, this topic certainly deserves future speculation. In fact, the choice (1.15) refers to somewhat special cases, which correspond to symmetric-type collisions consistent with the geometries of lattice gas dynamics. However, one should be able to define the collisional operator in the most general case.

1.2.2   *The Discrete Boltzmann Equation for Gas Mixtures*

The theory of the discrete Boltzmann equation for gas mixtures has been developed in [7] starting form the pioneer work [6]. The modelling procedure follows the same lines needed to obtain the equation for a simple gas, which we have seen in the preceding section. The main difference consists in the fact that particles belonging to different species may interact with each other exchanging momentum, according to the requirement that the two species *mix* themselves.

Paper [6] and afterwards [7] have shown that one can keep for all gas components the same discretization of the velocity directions, i.e. the vectors $\mathbf{e}_i$, and choose properly the velocity moduli in order to preserve momentum and energy, also in the interaction between particles of different species.

In order to understand this point, consider a binary mixture of two gases with masses m_1 and m_2 and let $\mu = m_1/m_2$ be the mass ratio. Assuming, for simplicity, that the gas may only undergo binary collisions, consider the velocity discretization

$$\mathbf{v}_i^1 = \begin{cases} ce_i & \text{if } i = 1, \ldots, n^* < n \\ \alpha ce_i & \text{if } i = n^* + 1, \ldots, n \, , \end{cases} \qquad (1.16a)$$

for the first gas and

$$\mathbf{v}_i^2 = \begin{cases} \mu ce_i & \text{if } i = 1, \ldots, n^* < n \\ \alpha \mu ce_i & \text{if } i = n^* + 1, \ldots, n \, , \end{cases} \qquad (1.16b)$$

for the second gas. The selection of similar discretizations for both gas components, i.e. $\mathbf{v}_i^2 = \mu \mathbf{v}_i^1$, $\forall i = 1, \ldots, n$ is motivated by two fundamental arguments. The proportionality of the two models assures the same collisional scheme within each gas component, while choosing the constant of proportionality equal to μ allows interactive collisions among particles of the two species. In fact, if the collision of the type

$$(\mathbf{v}_i^1, \mathbf{v}_j^1) \longleftrightarrow (\mathbf{v}_h^1, \mathbf{v}_k^1) \qquad i, j, h, k = 1, \ldots, n$$

between particles of the first gas is admissible, i.e. preserves momentum and energy, then also

$$(\mathbf{v}_i^2, \mathbf{v}_j^2) \longleftrightarrow (\mathbf{v}_h^2, \mathbf{v}_k^2) \qquad i, j, h, k = 1, \ldots, n$$

and

$$(\mathbf{v}_i^1, \mathbf{v}_j^2) \longleftrightarrow (\mathbf{v}_h^1, \mathbf{v}_k^2) \qquad i, j, h, k = 1, \ldots, n \qquad (1.17)$$

$$
\text{simple gas} \left\{
\begin{array}{ccc}
ce_i & \longleftrightarrow & \alpha ce_j \\
\updownarrow & & \updownarrow \\
ce_h & \longleftrightarrow & \alpha ce_k
\end{array}
\right.
$$

$$
\updownarrow \qquad\qquad \updownarrow \qquad\qquad \text{binary mixture}
$$

$$
\text{simple gas} \left\{
\begin{array}{ccc}
\mu ce_i & \longleftrightarrow & \mu\alpha ce_j \\
\updownarrow & & \updownarrow \\
\mu ce_h & \longleftrightarrow & \mu\alpha ce_k
\end{array}
\right.
$$

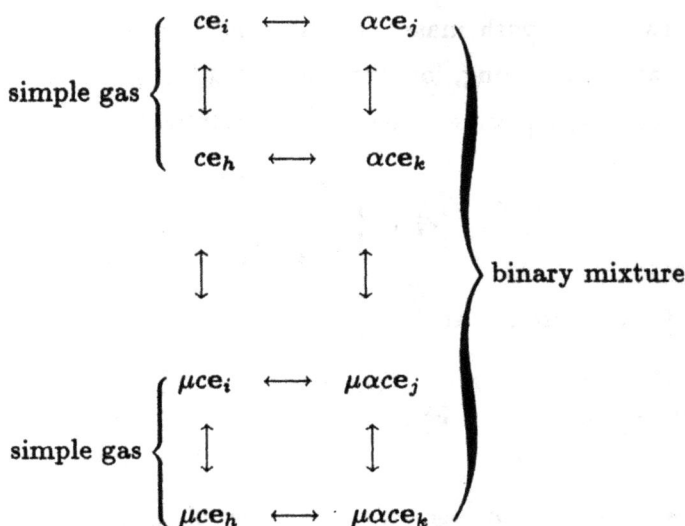

Fig. 1.1 — *Collision mechanics with two velocity moduli:*
one-component gas and binary mixture.

are admissible, i.e. the same type of conservation holds. Figure 1.1 visualize this aspect and indicates all conceivable collisions.

If $i = k$ and $j = h$ the interactive collisions (1.17) are defined exchange interactive collisions.

This type of procedure can be straightforwardly generalized to multicomponent gas mixtures. The formal expression of the discrete Boltzmann equation for a gas mixture of P gases with molecular masses

m_p , $p = 1, \ldots, P$ and velocities

$$\mathbf{v}_i^p = \mu_p \mathbf{v}_i^1 \quad , \quad i = 1, \ldots, n \quad , \quad \mu_p = \frac{m_1}{m_p}$$

is

$$N_i^p = N_i^p(t, \mathbf{x}) : (t, \mathbf{x}) \in [0, \mathcal{T}] \times \mathbb{R}^d \longmapsto \mathbb{R}_+ \quad , \quad \begin{matrix} i = 1, \ldots, n \\ p = 1, \ldots, P \end{matrix}$$

$$(\frac{\partial}{\partial t} + \mathbf{v}_i^p \cdot \nabla_\mathbf{x}) N_i^p = J_i^p[\mathbf{N}] = \check{J}_i^p[\mathbf{N}^p] + \tilde{J}_i^p[\mathbf{N}] \quad , \tag{1.18}$$

where

$$\mathbf{N}^P = \{N_1^P, \ldots, N_n^P\} \in \mathbb{R}^n \quad ,$$
$$\mathbf{N} = \{N_1^1, \ldots, N_n^1, \ldots, N_1^P, \ldots, N_n^P\} \in \mathbb{R}^{nP} \tag{1.19}$$

and where, according to a generalization of the collision scheme of Fig. 1.1, the collision operators $\check{J}_i^p$ and $\tilde{J}_i^p$, which correspond, respectively, to collisions between particles of the same gas and to interactive collisions between particles of different gases, can be written as

$$\check{J}_i^p[\mathbf{N}^p] = \frac{1}{2} \sum_{jhk} A_{pi,pj}^{ph,pk} (N_h^p N_k^p - N_i^p N_j^p) \tag{1.20}$$

$$\tilde{J}_i^p[\mathbf{N}] = \frac{1}{2} \sum_{r \neq p} \sum_{jhk} A_{pi,rj}^{ph,rk} (N_h^p N_k^r - N_i^p N_j^r) \ . \tag{1.21}$$

Once more, the transition rates are related to the transition probability densities by the formula

$$A_{pi,rj}^{ph,rk} = S_{pr}|\mathbf{v}_i^p - \mathbf{v}_j^r|a_{pi,rj}^{ph,rk} , \qquad (1.22)$$

where the constants S_{pr} are the cross-sectional areas of the interaction between p and r particles and the transition probability densities $a_{pi,rj}^{ph,rk}$ must be such that microreversibility and indistinguishability holds

$$a_{pi,rj}^{ph,rk} = a_{rj,pi}^{ph,rk} = a_{pi,rj}^{rk,ph} = a_{rj,pi}^{rk,ph} \; ; \; a_{pi,rj}^{ph,rk} = a_{ph,rk}^{pi,rj} , \qquad (1.23)$$

with mass conservation in the collision

$$\sum_{h,k=1}^{n} a_{pi,rj}^{ph,rk} = 1 \qquad \forall i,j = 1,\ldots,n \qquad \forall p,r = 1,\ldots,P \qquad (1.24)$$

and such that $a_{pi,rj}^{ph,rk} = \frac{1}{q}$ for all q outputs of the type $(\mathbf{v}_h^p, \mathbf{v}_k^r)$ corresponding to the input $(\mathbf{v}_i^p, \mathbf{v}_j^r)$.

The literature of models of gas mixtures with multiple collisions is limited only to paper [14], which will not be used in the analysis of fluid-dynamic problems in this volume, therefore this topic is not dealt with here, as it needs to be regarded at a very preliminary stage.

1.3   Macroscopic Observables

Having in mind fluid-dynamic problems, it is useful to show how one can pass from the microscopic description, namely the number densities

N_i^p, to the macroscopic observables. In this section these definitions are given for a gas mixture with P species. It is plain that the case of a single gas is recovered by setting $P = 1$ and removing any dependence on the index p.

The microscopic state is obtained solving suitable mathematical problems, initial or initial-boundary value problems, related to the discrete Boltzmann equation, i.e. the set of evolution equations (1.18) for the densities N_i^P.

Then one can transfer these information from the microscopic scale to the macroscopic scale exploiting the definitions of macroscopic observables which can be obtained in a way similar to the one used in continuous kinetic theory, that is by moments of the number densities. The only difference is in the calculation of these moments. In fact in continuous kinetic theory the moments are computed by integration over the continuous set of velocities, while in discrete theory the integrals are substituted by sums.

Therefore we can define

- Numerical density of each p–species

$$\nu_p(t, \mathbf{x}) = \sum_{i=1}^{n} N_i^P(t, \mathbf{x}) \,, \tag{1.25}$$

- Mass density of each p–species

$$\rho_p(t, \mathbf{x}) = m_p \nu_p(t, \mathbf{x}) \,. \tag{1.26}$$

The corresponding quantities of the whole mixture are immediately obtained by summation over the p–index

– Total numerical density

$$\nu(t, \mathbf{x}) = \sum_{p=1}^{P} \nu_p(t, \mathbf{x}) \,, \qquad (1.27)$$

– Total mass density

$$\rho(t, \mathbf{x}) = \sum_{p=1}^{P} \rho_p(t, \mathbf{x}) \,. \qquad (1.28)$$

Further macroscopic observables can be derived as higher order moments of the number densities N_i^p. In details, if a quantity F_i^p, function of velocity, is associated to any number density N_i^p, its average value over the p–species is given by

$$\frac{1}{\nu_p} \sum_{i=1}^{n} F_i^p N_i^p \,, \qquad (1.29)$$

while if one refers to the whole mixture, then the average must be computed by summation also over all the p–species

$$\frac{1}{\nu} \sum_{p=1}^{P} \sum_{i=1}^{n} F_i^p N_i^p \,. \qquad (1.30)$$

Consequently one has

– Mean velocity (drift velocity) of each gas-component

$$\mathbf{u}_p(t, \mathbf{x}) = \frac{1}{\nu_p} \sum_{i=1}^{n} \mathbf{v}_i^p N_i^p(t, \mathbf{x}) \,, \qquad (1.31)$$

- Mean velocity of the whole mixture

$$\mathbf{u}(t, \mathbf{x}) = \frac{1}{\nu} \sum_{p=1}^{P} \sum_{i=1}^{n} \mathbf{v}_i^P N_i^P(t, \mathbf{x}) \ . \tag{1.32}$$

Moreover we can define in terms of the mass velocities

$$\mathbf{w}_i^P(t, \mathbf{x}) = \mathbf{v}_i^P - \mathbf{u}_p(t, \mathbf{x}) \ , \tag{1.33}$$

the following additional quantities

- Stress tensor

$$[\mathbf{P}](t, \mathbf{x}) = \{\mathcal{P}_{jk}\}_{j,k=1,\dots,d}(t, \mathbf{x})$$

$$= \sum_{p=1}^{P} m_p \sum_{i=1}^{n} \mathbf{w}_i^P(t, \mathbf{x}) \otimes \mathbf{w}_i^P(t, \mathbf{x}) N_i^P(t, \mathbf{x}) \ , \tag{1.34}$$

where $\otimes$ denotes the tensor product and d the dimension of the space,

- Hydrodynamic pressure

$$\Pi(t, \mathbf{x}) = \frac{1}{d} \sum_{j=1}^{d} \mathcal{P}_{jj}(t, \mathbf{x}) \ , \tag{1.35}$$

and from the law of perfect gases ($\Pi(t, \mathbf{x}) = k_B \nu(t, \mathbf{x}) T(t, \mathbf{x})$, k_B being the Boltzmann constant)

- Temperature

$$T(t, \mathbf{x}) = \frac{1}{d k_B \nu(t, \mathbf{x})} \sum_{p=1}^{P} m_p \sum_{i=1}^{n} |\mathbf{w}_i^P(t, \mathbf{x})|^2 N_i^P(t, \mathbf{x}) \ , \tag{1.36}$$

which, recalling (1.33), can also be written as

$$T(t, \mathbf{x}) = \frac{1}{dk_B\nu} \sum_{p=1}^{P} m_p \sum_{i=1}^{n} \left(|\mathbf{v}_i^p|^2 N_i^p - |\mathbf{u}_p|^2 \nu_p \right) \ .$$

Furthermore we will also use the following definitions

- Heat flux

$$\mathbf{q}(t, \mathbf{x}) = \frac{1}{2} \sum_{p=1}^{P} m_p \sum_{i=1}^{n} \mathbf{w}_i^p(t, \mathbf{x}) |\mathbf{w}_i^p(t, \mathbf{x})|^2 N_i^p(t, \mathbf{x}) , \qquad (1.37)$$

- Entropy

$$\mathcal{S}(t, \mathbf{x}) = -k_B \sum_{p=1}^{P} \sum_{i=1}^{n} N_i^p(t, \mathbf{x}) \log N_i^p(t, \mathbf{x}) . \qquad (1.38)$$

The physical quantities which have been defined above are referred to the whole mixture. Obviously such quantities can be analogously written for each gas-species simply removing the summation over the p–index and substituting ν with ν_p.

Finally, we remark that in the case of a model in which any gas component can move with only one velocity modulus, say c_p, the temperature of the p–species can be written as

$$T_p(t, \mathbf{x}) = \frac{m_p}{dk_B} (c_p^2 - |\mathbf{u}_p|^2) \qquad (1.39)$$

and hence it is not an independent macroscopic observable.

1.4 Thermodynamic Equilibrium and Conservation Equations

The analysis of the properties of thermodynamic equilibrium for the discrete Boltzmann equation is essentially due to Gatignol [1]. The main results are already contained in the Lecture Notes [1] and are furtherly developed in various papers of the same author and co-workers [9,15]. Additional studies are due to Shizuta and Kawashima [12].

A detailed definition of thermodynamic equilibrium states is of crucial importance, as we shall see, for the analysis of several problems in fluid-dynamics. In fact, several flows, e.g. flows related to shock waves, in ducts or around convex bodies, can be regarded as deviations, either small or large, from suitable equilibrium states.

The content of this section is organized in two parts, which deal, respectively, with a monoatomic gas and with gas mixtures.

1.4.1 *Analysis for a One-Component Gas*

As already explained in Section 1.2.1, among all interactions one only considers those which have the same total mass, momentum and energy before and after the collision. There are then some functions of the set of discretized velocities which remain unchanged during the admissible collisions. Hence, the following definition will be useful

Definition 1.1: *A vector $\Psi \in \mathbf{R}^n$, with components Ψ_i functions of the velocity discretization, is a collision invariant if and only if*

$$< \Psi, \mathbf{J}^{(s)} >= 0 \qquad for \quad s = 2,3 \ , \qquad (1.40)$$

where the inner product is defined as

$$< \mathbf{F}, \mathbf{G} >= \sum_{i=1}^{n} F_i G_i \quad , \qquad \mathbf{F}, \mathbf{G} \in \mathbf{R}^n$$

and where $\mathbf{J}^{(s)} = \mathbf{J}^{(s)}[\mathbf{N}] = \{J_i^{(s)}[\mathbf{N}]\} \in \mathbf{R}^n$. The index s refers to the part of the collision operator which takes care of the binary ($s = 2$) and of the triple ($s = 3$) collisions.

The space of collision invariants, which is a linear subspace of $\mathbf{R}^n$, is denoted by $\mathcal{M}$. The dimension of $\mathcal{M}$, i.e. the number of linearly independent collision invariants, is denoted by δ, while $\mathbf{\Phi}^{(\chi)}$, $\chi = 1, \ldots, \delta$ denotes an orthogonal basis of $\mathcal{M}$.

Remark 1.1: *It is a matter of technical calculations, which exploits the reversibility and permutability properties (1.6–1.7) of the transition rates, to prove that* $\mathbf{\Psi}$ *is a collision invariant if and only if it satisfies the relations*

$$\Psi_i + \Psi_j = \Psi_h + \Psi_k , \tag{1.41a}$$

for every admissible binary collision $(\mathbf{v}_i, \mathbf{v}_j) \longleftrightarrow (\mathbf{v}_h, \mathbf{v}_k)$ and

$$\Psi_i + \Psi_j + \Psi_\ell = \Psi_g + \Psi_h + \Psi_k , \tag{1.41b}$$

for every admissible triple collision $(\mathbf{v}_i, \mathbf{v}_j, \mathbf{v}_\ell) \longleftrightarrow (\mathbf{v}_g, \mathbf{v}_h, \mathbf{v}_k)$.

Thanks to (1.10), in the more general case of a spatial model, the vectors

$$\boldsymbol{\Psi}^{(1)} = \{1, \ldots, 1\}$$

$$\boldsymbol{\Psi}^{(2)} = \frac{1}{c}\{v_{1x}, \ldots, v_{nx}\}$$

$$\boldsymbol{\Psi}^{(3)} = \frac{1}{c}\{v_{1y}, \ldots, v_{ny}\} \qquad (1.42)$$

$$\boldsymbol{\Psi}^{(4)} = \frac{1}{c}\{v_{1z}, \ldots, v_{nz}\}$$

$$\boldsymbol{\Psi}^{(5)} = \frac{1}{c^2}\{|\mathbf{v}_1|^2, \ldots, |\mathbf{v}_n|^2\}$$

are collisional invariants.

In general, a good model should be characterized by the correct space of collision invariants corresponding, as for the continuous Boltzmann equation, to conservation of mass, momentum and energy. Moreover, the model should show stability properties in the perturbation of the equilibrium state. Models which satisfy such properties are called *regular* [12,13]. The dimension of the space of collision invariants can be computed by a detailed analysis of the collision mechanics [16] and consequently one has, as we shall see, the conservation equations for the hydrodynamic moments, defined by

$$W_\chi = W_\chi(t, \mathbf{x}) = < \boldsymbol{\Phi}^{(\chi)}, \mathbf{N} > \; , \quad \chi = 1, \ldots, \delta \; . \qquad (1.43)$$

If the model is regular, these quantities are one-to-one related with the independent macroscopic observables.

Another important definition to introduce is the one regarding the equilibrium state

Definition 1.2:   *A vector $\widehat{\mathbf{N}} = \{\widehat{N_i}\} \in \mathbb{R}^n$ is a Maxwellian if $\widehat{\mathbf{N}}$ is non negative, i.e. $\widehat{N_i} \geq 0$, and if $J_i[\widehat{\mathbf{N}}] = 0$ for $i = 1,\ldots,n$.*

One of the crucial point is then how to relate the Maxwellian state $\widehat{\mathbf{N}}$, i.e. the parameters characterizing $\widehat{\mathbf{N}}$, to the macroscopic observables at equilibrium.

To this aim, we recall the result obtained by Gatignol [1] who proved that for a simple gas with binary collisions only the H–type functional

$$H = \sum_{i=1}^{n} N_i \log N_i \tag{1.44}$$

has the property that

$$\frac{dH}{dt} \leq 0 , \tag{1.45}$$

where the equality sign holds in the Maxwellian state. Moreover, at equilibrium, the following relations are equivalent

i) $\dfrac{dH}{dt} = 0$,

ii) $\log \widehat{\mathbf{N}} = \{\log \widehat{N_1},\ldots,\log \widehat{N_n}\} \in \mathcal{M}$,

iii) $\mathbf{J}[\widehat{\mathbf{N}}] = \mathbf{0}$.

The generalization of this result to the discrete Boltzmann equation with both binary and triple collisions is technical [8].

This result can be used to recover the detailed expression of the densities at equilibrium. In fact, since $\log \widehat{\mathbf{N}}$ is a collision invariant, one has

$$\log \widehat{\mathbf{N}} = \sum_{\chi=1}^{\delta} c_\chi \mathbf{\Phi}^{(\chi)} , \qquad (1.46)$$

and consequently

$$\widehat{\mathbf{N}} = \exp \left[\sum_{\chi=1}^{\delta} c_\chi \mathbf{\Phi}^{(\chi)} \right] , \qquad (1.47)$$

where, of course, $e^{\mathbf{f}}$ stands for the vector $\{e^{f_1}, \ldots, e^{f_n}\} \in \mathbb{R}^n$. In particular, the terms c_χ can be regarded as constant quantities in the case of absolute Maxwellians or functions of time and space in the case of local Maxwellians.

If the model is regular, the hydrodynamic moments W_χ are one-to-one related with the independent macroscopic observables. If we prove that the Maxwellian parameters c_χ are one-to-one related with the hydrodynamic moments at equilibrium $\widehat{W}_\chi$, we have that they are also one-to-one related with the macroscopic observables at equilibrium.

In fact, using (1.43) and (1.47), one can write the elements of the jacobian of the map

$$c_\chi \longmapsto \widehat{W}_\chi \qquad (1.48)$$

as

$$\frac{\partial \widehat{W}_\chi}{\partial c_\xi} = < \frac{\partial \widehat{\mathbf{N}}}{\partial c_\xi}, \mathbf{\Phi}^{(\chi)} > = < \text{Diag}[\widehat{N}_i] \, \mathbf{\Phi}^{(\xi)}, \mathbf{\Phi}^{(\chi)} > = \sum_{i=1}^{n} \widehat{N}_i \Phi_i^{(\xi)} \Phi_i^{(\chi)} .$$

Hence the jacobian defines a quadratic form which is positive definite. In fact, multiplying this equation by the components of the vector $\mathbf{X} = \{X_1, \ldots, X_\delta\} \in \mathbb{R}^\delta$ and summing over the indexes ξ and χ leads to

$$\sum_{\xi,\chi=1}^{\delta} X_\chi \frac{\partial \widehat{W}_\chi}{\partial c_\xi} X_\xi = \sum_{i=1}^{n} \widehat{N}_i \left(\sum_{\chi=1}^{\delta} \Phi_i^{(\chi)} X_\chi \right)^2 > 0 \ ,$$

because, from (1.47), the Maxwellian densities $\widehat{N}_i$ are all strictly positive.

Unfortunaly, the map (1.48) cannot be explicitly inverted, unless for some simple models or in simple fluid-dynamic flows.

Assume now that the dimension δ of the space of collisional invariants for a spatial model of a single gas component is equal to 5, corresponding to conservation of mass, of the three components of momentum and of energy. It can be shown that the set of collisional invariants $\mathbf{\Psi}^{(\chi)}$ defined in Eq.(1.42) is a basis for $\mathcal{M}$.

One can then re-write the Maxwellians using this alternative basis, instead of the orthogonal one, so that (1.46) is replaced by

$$\log \widehat{N}_i = c_\nu + c_x \frac{v_{ix}}{c} + c_y \frac{v_{iy}}{c} + c_z \frac{v_{iz}}{c} + c_T \frac{|\mathbf{v}_i|^2}{c^2}$$

and hence the Maxwellian densities can be written as

$$\widehat{N}_i = A \exp \left[c_x \frac{v_{ix}}{c} + c_y \frac{v_{iy}}{c} + c_z \frac{v_{iz}}{c} + c_T \frac{|\mathbf{v}_i|^2}{c^2} \right] \qquad (1.49)$$

where $A = e^{c_\nu}$. Equation (1.49) holds even if the model is planar, but with $c_z = 0$, of course. Furthermore, if the temperature is not an independent observable, as in the case of models with a single velocity modulus (see the end of Section 1.3), Eq.(1.49) holds with $c_T = 0$.

Since both $\mathbf{\Psi}^{(\chi)}$ in Eq.(1.42) and $\mathbf{\Phi}^{(\chi)}$ are basis of $\mathcal{M}$ there is a one-to-one relation between c_χ, $\chi = 1, \ldots, 5$ and A, c_x, c_y, c_z, c_T. Therefore, according to (1.48), the map which relates $\widehat{W}_\chi$, $\chi = 1, \ldots, 5$ and A, c_x, c_y, c_z, c_T is again one-to-one.

If, furthermore, the model is such that any velocity has its opposite value, as it will be for all the models dealt with in the present book, then one has that absolute Maxwellians, i.e. Maxwellians with vanishing drift velocity, correspond to setting $c_x = c_y = c_z = 0$ in (1.49). In fact, if $c_x = c_y = c_z = 0$, then, for instance,

$$\widehat{\nu u_x} = A \sum_{i=1}^{n} v_{ix} e^{c_T |\mathbf{v}_i|^2 / c^2}$$

$$= A \sum_{i:v_{ix}>0} v_{ix} e^{c_T |\mathbf{v}_i|^2 / c^2} + A \sum_{i:v_{ix}<0} -|v_{ix}| e^{c_T |\mathbf{v}_i|^2 / c^2} = 0 \ .$$

In the same way one can also prove that $u_y = u_z = 0$. The reverse is proved by remembering that the map between $\{\nu, u_x, u_y, u_z, T\}$ and $\{A, c_x, c_y, c_z, c_T\}$ is one-to-one.

In order to relate the solution of the discrete Boltzmann equation to the solution of hydrodynamics, it is useful to derive the conservation equations for the moments W_χ.

In particular, the conservation equations can be obtained, first rewriting the kinetic equations in a suitable vector form

$$\frac{\partial}{\partial t} \mathbf{N} + [\mathbf{V}] \mathbf{N} = \mathbf{J}[\mathbf{N}] \ , \tag{1.50}$$

where $[\mathbf{V}]$ is the square diagonal matrix (operator) of rank n

$$[\mathbf{V}] = \mathrm{diag}\{\mathbf{v}_1 \cdot \nabla_{\mathbf{x}}, \ldots, \mathbf{v}_n \cdot \nabla_{\mathbf{x}}\} \ . \tag{1.51}$$

Taking the scalar product of the collisional invariant $\mathbf{\Phi}^{(x)}$ with (1.50) yields

$$< \mathbf{\Phi}^{(x)}, \frac{\partial \mathbf{N}}{\partial t} + [\mathbf{V}]\mathbf{N} > = < \mathbf{\Phi}^{(x)}, \mathbf{J}[\mathbf{N}] > = 0 \ , \quad \chi = 1, \ldots, \delta \ . \tag{1.52}$$

The conservation equations can then be written as

$$\frac{\partial W_\chi}{\partial t} + \mathcal{F}_\chi[\mathbf{N}] = 0 \ , \quad \chi = 1, \ldots, \delta \ , \tag{1.53}$$

where

$$\mathcal{F}_\chi[\mathbf{N}] = < \mathbf{\Phi}^{(x)}, [\mathbf{V}]\mathbf{N} > = \sum_{i=1}^{n} \Phi_i^{(x)} \mathbf{v}_i \cdot \nabla_{\mathbf{x}} N_i \ . \tag{1.54}$$

The system of partial differential equation defined in (1.53) is, generally, not closed, as the terms $\mathcal{F}_\chi$ are not only functions of W_χ , $\chi = 1, \ldots, \delta$, but also of all N_i.

A discrete velocity model of the Boltzmann equation admits a number of conservation equations equal to the dimension δ of the space of collision invariants and, therefore, if the model is non regular, there are some conservation equations which have no physical counterpart.

In general, given a discrete velocity model, it is always possible to find a system of n variables of the type

$$W_\chi = W_\chi[\mathbf{N}](t, \mathbf{x}) = \begin{cases} < \mathbf{\Phi}^{(x)}, \mathbf{N} > & \text{if } \chi = 1, \ldots, \delta \\ < \tilde{\mathbf{\Phi}}^{(x)}, \mathbf{N} > & \text{if } \chi = \delta + 1, \ldots, n \ , \end{cases} \tag{1.55}$$

where $\widetilde{\Phi}^{(\chi)}$ is a basis of the space $\mathcal{M}^{\perp}$ (orthogonal complement of $\mathcal{M}$). Hence, for $\chi = 1, \ldots, \delta$ the terms W_{χ} represents the physical hydrodynamic moments already defined in (1.43), and for $\chi = \delta + 1, \ldots, n$ some additional variable functions of $\mathbf{N}$.

Then the kinetic equations (1.50) yield

$$\frac{\partial W_{\chi}}{\partial t} + \mathcal{F}_{\chi}[\mathbf{W}] = Q_{\chi}[\mathbf{W}] \ , \quad \chi = 1, \ldots, n \ , \tag{1.56}$$

where $\mathbf{W} = \{W_{\chi}\} \in \mathbb{R}^{n}$ and

$$\begin{aligned} Q_{\chi}[\mathbf{W}] &= 0 \quad \text{if} \ \ 1 \leq \chi \leq \delta \\ Q_{\chi}[\mathbf{W}] &= < \widetilde{\Phi}^{(\chi)}, \mathbf{J}[\mathbf{N}] > \neq 0 \quad \text{if} \ \ \delta + 1 \leq \chi \leq n \ . \end{aligned} \tag{1.57}$$

Note that the first δ equations do not define a closed system as the terms W_{χ}, $\chi = 1, \ldots, \delta$ are also functions of W_{χ}, $\chi = \delta + 1, \ldots, n$.

The system of the first δ equations is closed, however, at equilibrium, $\mathbf{N} = \widehat{\mathbf{N}}$, and generate the Euler equations linked to a given discrete velocity model of the Boltzmann equation. Some examples will be given in Chapter 2.

1.4.2 Analysis for Gas Mixtures

The characterization of the equilibrium Maxwellian state for gas mixtures follows the same method used for a simple gas. One needs first to define the space of the collision invariants and then to generalize to gas mixtures the result of the analysis of the preceding section.

The definition of collisional invariant is a straightforward generalization of Definition 1.1.

Definition 1.3:   *A vector*

$$\Psi = \{\Psi_i^p\} = \{\Psi_1^1, \ldots, \Psi_n^1, \ldots, \Psi_1^P, \ldots, \Psi_n^P\} \in \mathbf{R}^{nP}$$

where the components are functions of the velocity discretization is a
collision invariant if and only if $< \Psi, \mathbf{J} > = 0$, *where* $\mathbf{J} = \{J_i^p[\mathbf{N}]\} \in$
$\mathbf{R}^{nP}$ *and the inner product is defined as*

$$< \mathbf{F}, \mathbf{G} > = \sum_{p=1}^{P} \sum_{i=1}^{n} F_i^p G_i^p \quad , \quad \mathbf{F}, \mathbf{G} \in \mathbf{R}^{nP} .$$

Again by technical calculation one can prove that Ψ is a collisional
invariant iff

$$\Psi_i^p + \Psi_j^r = \Psi_h^p + \Psi_k^r \tag{1.58}$$

for every admissible collision $(\mathbf{v}_i^p, \mathbf{v}_j^r) \longleftrightarrow (\mathbf{v}_h^p, \mathbf{v}_k^r)$.

In the case of gas mixtures, the dimension of the space $\mathcal{M}$ of the
collision invariants, which is now a linear subspace of $\mathbf{R}^{nP}$, is larger
that the one of a simple gas. In fact, one has conservation of masses
separately. According to [7] the dimension δ of the space $\mathcal{M}$ for the gas
mixture can be related to the dimension $\bar{\delta}$ of the space $\widetilde{\mathcal{M}}$ relative to
the single gas model, to the number of species P and to the number $\tilde{n}$ of
velocity moduli used, per each species, in the discretization, as follows

$$\delta = \bar{\delta} + (P - 1)\tilde{n} , \tag{1.59}$$

for $\tilde{n} \le 2$.

Having this result in mind (the proof is a matter of technical calculations which are not reported here), one can develop an analysis of the equilibrium thermodynamic state analogous to the one given for a simple gas. In particular, a vector $\widehat{\mathbf{N}} = \{\widehat{N}_i^p\} \in \mathbb{R}^{nP}$ is a Maxwellian if $\widehat{N}_i^p \ge 0$ and $J_i^p[\widehat{\mathbf{N}}] = 0$, $\forall i = 1, \ldots, n$ and $p = 1, \ldots, P$.

The H–functional, defined as

$$H = \sum_{p=1}^{P} \sum_{i=1}^{n} N_i^p \log N_i^p \ , \tag{1.60}$$

satisfies

$$\frac{dH}{dt} \le 0 \ , \tag{1.61}$$

where the equality sign holds in equilibrium [7].

Also in this case, the equilibrium conditions are characterized by the equivalence of the following conditions

i) $\dfrac{dH}{dt} = 0$,

ii) $\log \widehat{\mathbf{N}} = \{\log \widehat{N}_1^1, \ldots, \log \widehat{N}_n^1, \ldots, \log \widehat{N}_1^P, \ldots, \log \widehat{N}_n^P\} \in \mathcal{M}$, (1.62)

iii) $\mathbf{J}[\widehat{\mathbf{N}}] = \mathbf{0}$.

The Maxwellian densities, for each species, then take the form

$$\widehat{N}_i^p = \exp\left[\sum_{\chi=1}^{\delta} c_\chi \Phi_i^{p(\chi)}\right] \ , \tag{1.63}$$

where, in general, $c_\chi = c_\chi(t, \mathbf{x})$ and where the collision invariants $\mathbf{\Phi}^{(x)} = \{\Phi_i^{p(x)}\} \in \mathbf{R}^{nP}$ define the components of an orthogonal basis of the linear space $\mathcal{M}$.

The Maxwellian parameters c_χ are again one-to-one related with the hydrodynamic moments $\widehat{W}_\chi$ with the same implications which we have seen in the preceding section.

1.5 Plan of the Book

A detailed plan of the content of this book can now be given in conclusion of this chapter. As already mentioned, Chapter 2 provides the description of several models of the discrete Boltzmann equation in view of the fluid-dynamic applications dealt with in several chapters of the book. These models are all derived according to the methodology given in Section 1.2. In particular, for each model, various details will be given with reference to the discretization of the velocity space, to the construction of the model and to the characterization of the Maxwellian equilibrium properties.

Chapter 3 deals with the solution of the discrete Boltzmann equation in unbounded domains. The first part of the chapter provides, with reference to the review papers [4,5], a survey of the mathematical result in the existing literature on the analysis of existence of solutions to the Cauchy problem for the discrete Boltzmann equation. The second part of the chapter deals with a presentation of various solution techniques: analytic, numerical and simulation schemes. This chapter can be regarded as the preliminary one to the content of the chapters which follow and which deal with several fluid-dynamic aspects of the discrete Boltzmann equation.

Chapter 4 deals with the analysis of shock wave phenomena described by means of the discrete Boltzmann equation either for a simple gas or for gas mixtures. This topic was introduced in the pioneer work by Broadwell [3] and afterwards developed by various authors. Among the various contributions one can cite Gatignol [17], Caflisch [18] and Kawashima and Matsumura [19] for a simple gas with binary collisions only. The same topic was developed, as we shall see, for gas mixtures [20,21] and for equations with multiple collisions [14,22]. Chapter 4 organizes the whole matter in a unified presentation which includes the description of the methodological aspects, qualitative analysis and quantitative results.

Chapter 5 deals with the analysis of fluid-dynamic problems in the presence of solid walls moving or at rest. The first step in the analysis of this type of problems is the statement of the boundary conditions on the boundary corresponding to the solid walls or, in the language of physicists, the analysis of the behaviour of the gas near the walls.

The first paper to deal in a systematic way with this topic is due to Gatignol [23], who introduced a matter afterwards developed and generalized in [24]. Once the boundary conditions have been stated, one can obtain the description of the flow patterns by solution of suitable initial-boundary (or boundary) value problems. This is the topic dealt with in Chapter 5 for various physically interesting problems. It needs to be mentioned that the mathematical aspects of this problem have the support of an existence theory, which provides qualitative analysis of the mathematical problem for a gas in a slab. The theory is developed by Kawashima [25] in the case of Maxwell equilibrium on both walls, by A. Pulvirenti [26] for a gas with specularly reflecting walls and in [27] for general boundary conditions of deterministic and stochastic type.

Finally, the last chapter deals with the analysis, modelling and shock wave phenomena of the discrete Boltzmann equation for chemically reacting gases [28–30]. After having introduced this topic, which is certainly of great physical interest, the chapter provides some ideas for the mathematical modelling of the discrete Boltzmann equation for a gas undergoing various types of chemical reactions and analyses some aspects related to the thermodynamic equilibrium referred to some specific models. The analysis of shock wave phenomena in the presence of chemical reactions is also dealt with as a conclusive application.

Each chapter, as already mentioned in the preface, provides suitable indications on open problems for applied mathematicians and fluid-dynamicists interested in nonlinear kinetic theory and its applications and suggests developments useful to a deeper understanding of the topic. Some of these indications can be regarded as suggestions of future research activities.

References

[1] R. Gatignol, **Théorie Cinétique des Gaz a Répartition Discrète de Vitesses**, *Lecture Notes in Phys.* No.**36**, Springer-Verlag, Berlin, New York, 1975.

[2] I.E. Broadwell, "Shock structure in a simple discrete velocity gas", *Phys. Fluids*, **7**, 1964, p.1243.

[3] I.E. Broadwell, "Study of rarefied shear flow by the discrete velocity method", *J. Fluid Mech.*, **19**, 1964, p.401.

[4] T. Platkowski and R. Illner, "Discrete velocity models of the Boltzmann equation: A survey on the mathematical aspects of the theory", *SIAM Review*, **30**, 1988, p.213.

[5] N. Bellomo and T. Gustafsson, "The discrete Boltzmann equation: A review of the mathematical aspects of the initial and initial-boundary value problem", *Review Math. Phys.*, to appear in 1991.

[6] N. Bellomo and L. M. de Socio, "The discrete Boltzmann equation

for gas mixtures. A regular space model and shock wave problem", *Mech. Res. Comm.*, **10**, 1983, p.233.

[7] E. Longo and R. Monaco, "On the thermodynamics of the discrete models of the Boltzmann equation for gas mixtures", *Transp. Theory Statist. Phys.*, **17**, 1988, p.423.

[8] E. Longo and R. Monaco, "On the discrete kinetic theory with multiple collisions: A plane six velocity model and unsteady Couette flow", in **Rarefied Gas Dynamics: Theoretical and Computational Techniques**, Eds. E.P. Muntz, D.H. Campbell and D.P. Weaver, AIAA Publ. vol.**118**, Washington, 1989, p.118.

[9] R. Gatignol and F. Coulouvrat, "Description hydrodinamique d'un gaz en théorie cinétique discrète", *Comp. Rend. Acad. Sci. Paris*, *I*, **306**, p.169 and p.393.

[10] N. Bellomo and S. Kawashima, "The discrete Boltzmann equation with multiple collisions: Global existence and stability for the initial value problem", *J. Math. Phys.*, **31**, 1990, p.245.

[11] H. Cabannes, "Etude de la propagation des ondes dans un gaz a quatorze vitesses", *J. de Mecanique*, **14**, 1975, p.705.

[12] Y. Shizuta and S. Kawashima, "The regular discrete models of the Boltzmann equation", *J. Math. Kyoto Univ.*, **27**, 1987, p.131.

[13] Y. Shizuta and S. Kawashima, "The regularity of discrete models of the Boltzmann equation", *Proc. Japan Acad. Ser. A*, **61**, 1985, p.252.

[14] M. Pandolfi Bianchi, "Modelling and nonlinear shock waves for binary gas mixtures by the discrete Boltzmann equation with multiple collisions", *Transp. Theory Statist. Phys.*, to appear in 1991.

[15] P. Chauvat and R. Gatignol, "Macroscopic variables in discrete kinetic theory", in **Discrete Models of Fluid Dynamics**, *Advances in Mathematics for Applied Sciences* vol.2, Ed. A. Alves, World Scientific, London, Singapore, 1991, p.1.

[16] P. Chauvat, "Summational invariants in discrete kinetic theory with multiple collisions", *Mech. Res. Comm.*, **18**, 1991, p.11.

[17] R. Gatignol, "Kinetic theory for a discrete velocity gas and application to the shock structure", *Phys. Fluids*, **18**, 1975, p.153.

[18] R. Caflisch, "Navier-Stokes and Boltzmann shock profiles for a model of gasdynamics", *Comm. Pure Appl. Math.*, **32**, 1979, p.521.

[19] S. Kawashima and A. Matzumura, "Asymptotic stability of travelling wave solutions of systems for one dimensional gas motion", *Comm. Math. Phys.*, **101**, 1985, p.97.

[20] R. Monaco, "Shock waves propagation in gas mixtures by a discrete velocity model of the Boltzmann equation", *Acta Mechanica*, **55**, 1985, p.239.

[21] R. Monaco, M. Pandolfi Bianchi and T. Platkowski, "Shock-waves formation by the discrete Boltzmann equation for binary gas mixtures", *Acta Mechanica*, **84**, 1990, p.175.

[22] N. Bellomo and E. Longo, "Shock wave profiles in one dimension by the discrete Boltzmann equation with multiple collisions", in **Waves and Stability in Continuous Media**, *Advances in Mathematics for Applied Sciences* vol.4, Ed. S. Rionero, World Scientific, London, Singapore, p.22.

[23] R. Gatignol, "Kinetic theory boundary conditions for discrete velocity gases", *Phys. Fluids*, **20**, 1977, p.2022.

[24] L. Preziosi and E. Longo, "On the decomposition of domains in non linear discrete kinetic theory", in **Discrete Models of Fluid Dynamics**, *Advances in Mathematics for Applied Sciences* vol.2, Ed. A. Alves, World Scientific, London, Singapore, p.144.

[25] S. Kawashima, "Global existence and asymptotic behaviour of solutions to the mixed problem for the discrete Boltzmann equation", Internal Report, 1990.

[26] A. Pulvirenti, "Global solution to the initial-boundary value problem for the discrete Boltzmann equation", *Transp. Theory Statist. Phys.*, to appear in 1991.

[27] Z. Brzeźniak, F. Flandoli and L. Preziosi, "On the discrete Boltzmann equation with multiple collisions", *Stab. and Appl. Anal. in Continuous Media*, to appear in 1991.

[28] E. Gabetta and R. Monaco, "On the modelling of the discrete Boltzmann equation for gases with bi-molecular chemical reactions", in **Discrete Models of Fluid Dynamics**, *Advances in Mathematics for Applied Sciences* vol.2, Ed. A. Alves, World Scientific, London, Singapore, 1991, p.22.

[29] R. Monaco and M. Pandolfi Bianchi, "A discrete velocity model for gases with chemical reactions of dissociation and recombination", in **Advances in Kinetic Theory and Continuum Mechanics**, Eds. R. Gatignol and Soubbaramayer, Springer-Verlag, Berlin, 1991, p.169.

[30] R. Monaco and M. Pandolfi Bianchi, "Shock wave onset with chemical dissociation by the discrete Boltzmann equation", in **Rarefied Gas Dynamics**, Ed. A. Beylich, VCH-Verlag, Weinheim, New

York, 1991, p.862.

CHAPTER 2

SOME DISCRETE VELOCITY MODELS

Several models of the discrete Boltzmann equation are described, in this chapter, according to the modelling methodology given in Chapter 1. Particular attention is paid to those models which will be used in the fluid-dynamic applications developed in the proceeding chapters.

In details, the first section of this chapter deals with planar models both in the case of binary and triple collisions, the second section with spatial models, like the ones due to Broadwell, Cabannes and Shizuta and Kawashima.

The presentation is organized providing first the discretization of the velocity space and the collision dynamics, then the kinetic equations are derived and the analysis of the thermodynamic equilibrium follow.

The aim of this chapter is to provide the description both of the models used in the analysis of fluid-dynamic problems, developed in Chapters 3 through 5, and of the models which may be used by the reader for future developments.

The content of this chapter is also somewhat referred to Chapter 6. In fact, the analysis of chemically reacting gases, which is still at a preliminary stage, can be furtherly developed making proper use of the models described in this section.

2.1   Planar 2n–Velocity Models

Consider a one-component discrete velocity gas such that the particles can attain $2n$ velocities in the xy–plane. In particular, the velocity discretization is characterized by the following properties

1) $|\mathbf{v}_i| = c$,

2) $\mathbf{v}_i + \mathbf{v}_{i+n} = 0$;

3) $\mathbf{v}_i \cdot \mathbf{v}_{i+1} = c^2 \cos \frac{\pi}{n}$

for $i = 1, \ldots, 2n$, where the indexes are to be intended modulo $2n$ all through this section, that is $j = i + n > 2n \Rightarrow j = (i + n) - 2n$. Such a model is known in literature as the planar $2n$–velocity model [1]. Its geometry is visualized in Fig. 2.1.

If only binary collisions are dealt with, then the non-trivial admissible ones (where this term is used to denote those collisions which produce non-vanishing terms in the collisional operator) are the head-on collisions

$$(\mathbf{v}_i, \mathbf{v}_{i+n}) \longleftrightarrow (\mathbf{v}_j, \mathbf{v}_{j+n}) \qquad , \qquad \forall i, j = 1, \ldots, 2n \quad , \quad j \neq i \quad . \quad (2.1)$$

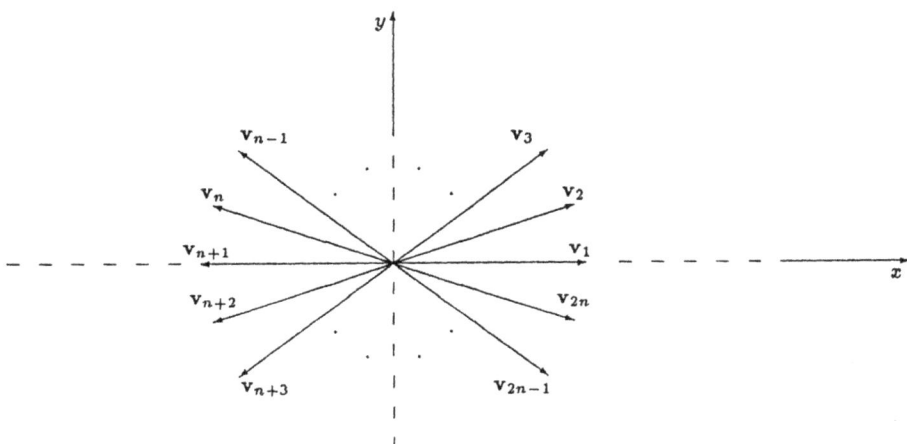

Fig. 2.1 — *Geometry of the 2n–planar model*

The kinetic equations can be derived starting from (2.1). In fact, the expression of the transition rates in the collisional term J_i is

$$A_{i,i+n}^{j,j+n} = S|\mathbf{v}_i - \mathbf{v}_{i+n}|a_{i,i+n}^{j,j+n} = 2cSa_{i,i+n}^{j,j+n} \quad . \tag{2.2}$$

Moreover, if all the velocity directions after collisions are assumed to be equally probable , then

$$\forall i \neq j \quad : \quad a_{i,i+n}^{j,j+n} = \frac{1}{n} \quad .$$

The kinetic equations assume the following form

$$\frac{\partial N_i}{\partial t} + c \left[\cos\left(\frac{i-1}{n}\pi\right) \frac{\partial N_i}{\partial x} + \sin\left(\frac{i-1}{n}\pi\right) \frac{\partial N_i}{\partial y} \right]$$

$$= \frac{2}{n}cS \sum_{\substack{j=1 \\ j \neq i}}^{n} (N_j N_{j+n} - N_i N_{i+n}) \tag{2.3}$$

$$= \frac{2}{n}cS \sum_{\ell=1}^{n-1} (N_{i+\ell} N_{i+\ell+n} - N_i N_{i+n}) \, ,$$

for $i = 1, \ldots, 2n$.

The collisional invariants obey the $n-1$ independent relations

$$\Psi_i + \Psi_{i+n} = \Psi_j + \Psi_{j+n} \qquad \forall j \neq i \, ,$$

then the dimension of the space of collisional invariant $\mathcal{M}$ is $\delta = 2n - (n-1) = n+1$, so that only the model with $n = 2$ has the correct dimension of the space of collisional invariants, related to conservation of mass and momentum. In fact, energy is not an independent quantity for models with only one velocity modulus (see Eq.(1.39)). If $n > 2$, then in the basis of $\mathcal{M}$ there are $n-2$ *spurious* collisional invariants, which have not a specific physical meaning, but are related to the geometrical configuration of the model.

2.1.1   *The 4–Velocity Model*

From the general $2n$–velocity model one can derive the well-known Gatignol model [2], letting $n = 2$. Its geometry is represented in Fig. 2.2.

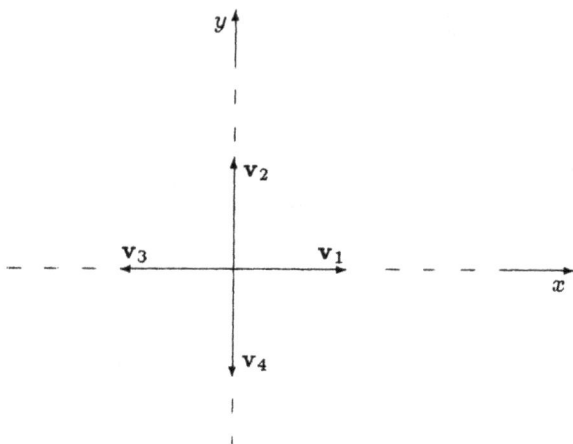

Fig. 2.2 — *Geometry of the discrete planar 4−velocity model*

The only admissible collision is

$$(\mathbf{v}_1, \mathbf{v}_3) \longleftrightarrow (\mathbf{v}_2, \mathbf{v}_4) \quad ,$$

with relative velocity $2c$ and probability density $1/2$, and hence the corresponding kinetic equations are

$$\frac{\partial N_1}{\partial t} + c\frac{\partial N_1}{\partial x} = cS(N_2 N_4 - N_1 N_3)$$

$$\frac{\partial N_2}{\partial t} + c\frac{\partial N_2}{\partial y} = cS(N_1 N_3 - N_2 N_4)$$

$$\frac{\partial N_3}{\partial t} - c\frac{\partial N_3}{\partial x} = cS(N_2 N_4 - N_1 N_3) \qquad (2.4)$$

$$\frac{\partial N_4}{\partial t} - c\frac{\partial N_4}{\partial y} = cS(N_1 N_3 - N_2 N_4) \ .$$

Since the collisional invariants obey the relation

$$\Psi_1 + \Psi_3 = \Psi_2 + \Psi_4 ,$$

there is a unique constraint among the four components of the collisional invariant. Thus, there are three independent collisional invariants. An orthogonal basis of the space $\mathcal{M}$ is

$$\Phi^{(1)} = \{1, 1, 1, 1\}$$
$$\Phi^{(2)} = \{1, 0, -1, 0\} \tag{2.5}$$
$$\Phi^{(3)} = \{0, 1, 0, -1\} .$$

Such collisional invariants correspond to conservation of mass and of the two components of momentum in the plane.

The thermodynamic equilibrium number densities can then be expressed, according to the analysis of Section 1.4, by

$$\widehat{N}_1 = \exp[c_1 + c_2]$$
$$\widehat{N}_2 = \exp[c_1 + c_3]$$
$$\widehat{N}_3 = \exp[c_1 - c_2] \tag{2.6}$$
$$\widehat{N}_4 = \exp[c_1 - c_3] ,$$

where the three parameters c_1, c_2, c_3 are related, at equilibrium, to the three independent macroscopic quantities

$$\widehat{\nu} = \widehat{N}_1 + \widehat{N}_2 + \widehat{N}_3 + \widehat{N}_4 = 2e^{c_1} (\cosh c_2 + \cosh c_3)$$
$$\widehat{u}_x = \frac{c}{\widehat{\nu}}(\widehat{N}_1 - \widehat{N}_3) = c\frac{\sinh c_2}{\cosh c_2 + \cosh c_3} \tag{2.7}$$
$$\widehat{u}_y = \frac{c}{\widehat{\nu}}(\widehat{N}_2 - \widehat{N}_4) = c\frac{\sinh c_3}{\cosh c_2 + \cosh c_3} .$$

The conservation equations for this model can be computed following the procedure presented in Section 1.4. In fact,

$$< \Phi^{(\chi)}, \left(\frac{\partial \mathbf{N}}{\partial t} + [\mathbf{V}]\mathbf{N} \right) >= 0 \quad , \quad \chi = 1, 2, 3$$

yields

$$\frac{\partial}{\partial t}(N_1 + N_2 + N_3 + N_4) + c\frac{\partial}{\partial x}(N_1 - N_3) + c\frac{\partial}{\partial y}(N_2 - N_4) = 0$$

$$\frac{\partial}{\partial t}(N_1 - N_3) + c\frac{\partial}{\partial x}(N_1 + N_3) = 0 \tag{2.8}$$

$$\frac{\partial}{\partial t}(N_2 - N_4) + c\frac{\partial}{\partial x}(N_2 + N_4) = 0 \,,$$

where, according to the definition of the macroscopic observables,

$$N_1 + N_2 + N_3 + N_4 = \nu \quad , \quad N_1 - N_3 = \nu\frac{u_x}{c} \quad , \quad N_2 - N_4 = \nu\frac{u_y}{c} \,,$$

but neither $N_1 + N_3$ nor $N_2 + N_4$ can be related to the density or the drift velocity. Hence, as explained in Section 1.4.1, the set of conservation equations is not closed.

In thermodynamic equilibrium, however, one has, from Eqs.(2.6–2.7), that

$$\widehat{N}_1 + \widehat{N}_3 = 2e^{c_1} \cosh c_2 = \frac{\widehat{\nu}}{2}\left(1 + \frac{\widehat{u}_x^2 - \widehat{u}_y^2}{c^2} \right)$$

$$\widehat{N}_2 + \widehat{N}_4 = 2e^{c_1} \cosh c_3 = \frac{\widehat{\nu}}{2}\left(1 - \frac{\widehat{u}_x^2 - \widehat{u}_y^2}{c^2} \right) \,,$$

so that Eqs.(2.8) re-write

$$\frac{\partial}{\partial t}\widehat{\nu} + \frac{\partial}{\partial x}(\widehat{\nu}\widehat{u}_x) + \frac{\partial}{\partial y}(\widehat{\nu}\widehat{u}_y) = 0$$

$$\frac{\partial}{\partial t}(\widehat{\nu}\widehat{u}_x) + \frac{\partial}{\partial x}[\widehat{\nu}\frac{c^2 + \widehat{u}_x^2 - \widehat{u}_y^2}{2}] = 0$$

$$\frac{\partial}{\partial t}(\widehat{\nu}\widehat{u}_y) + \frac{\partial}{\partial y}[\widehat{\nu}\frac{c^2 - \widehat{u}_x^2 + \widehat{u}_y^2}{2}] = 0 \ ,$$

which represent the Euler equations for the 4–velocity model and express the time-space macroscopic behaviour of the gas in equilibrium conditions.

2.1.2 *The 4–Velocity Model for Binary Gas Mixtures*

The $2n$–velocity regular planar model can be generalized to gases with an arbitrary number of components using the procedure outlined in Chapter 1. Details will be given only for the 4–velocity model for a binary gas mixture [3].

The geometry of the model is shown in Fig. 2.3.

The velocity moduli are

$$i = 1,\ldots,4 : \qquad |\mathbf{v}_i^1| = c \qquad\qquad |\mathbf{v}_i^2| = \mu c$$

where $\mu = m_1/m_2 < 1$, with m_1 and m_2 molecular masses of the lighter and heavier gases, respectively.

The collisional scheme is the following

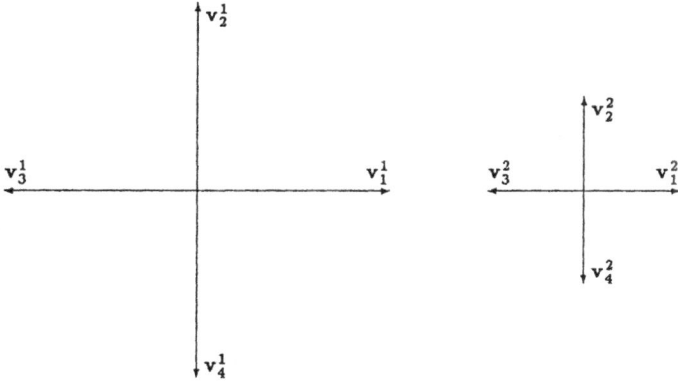

Fig. 2.3 — *Geometry of the 4–velocity model for a binary gas mixture*

a) Head-on collisions between particles of the lighter gas

$$(\mathbf{v}_1^1, \mathbf{v}_3^1) \longleftrightarrow (\mathbf{v}_2^1, \mathbf{v}_4^1)$$

with relative velocity $2c$ and probability density $1/2$.

b) Head-on collisions between particles of the heavier gas

$$(\mathbf{v}_1^2, \mathbf{v}_3^2) \longleftrightarrow (\mathbf{v}_2^2, \mathbf{v}_4^2)$$

with relative velocity $2\mu c$ and probability density $1/2$.

c) Head-on interactive collisions

$$(\mathbf{v}_1^1, \mathbf{v}_3^2) \longleftrightarrow (\mathbf{v}_3^1, \mathbf{v}_1^2) \longleftrightarrow (\mathbf{v}_2^1, \mathbf{v}_4^2) \longleftrightarrow (\mathbf{v}_4^1, \mathbf{v}_2^2)$$

with relative velocity $(1 + \mu)c$ and probability density $1/4$.

d) **Exchange interactive collisions**

$$i = 1, 3: \qquad (\mathbf{v}_i^1, \mathbf{v}_{i+1}^2) \longleftrightarrow (\mathbf{v}_{i+1}^1, \mathbf{v}_i^2)$$

$$(\mathbf{v}_i^1, \mathbf{v}_{i+3}^2) \longleftrightarrow (\mathbf{v}_{i+3}^1, \mathbf{v}_i^2)$$

with relative velocity $(1 + \mu^2)^{1/2}c$ and probability density $1/2$.

Let $\mathbf{e}_1 = \mathbf{i}$, $\mathbf{e}_2 = \mathbf{j}$, $\mathbf{e}_3 = -\mathbf{i}$, $\mathbf{e}_4 = -\mathbf{j}$, where $\mathbf{i}$ and $\mathbf{j}$ are the unit vectors of the orthogonal frame in the xy–plane. The kinetic equations are

$$\frac{\partial N_i^1}{\partial t} + c\mathbf{e}_i \cdot \nabla_{\mathbf{x}} N_i^1 = cS_{11}(N_{i+1}^1 N_{i+3}^1 - N_i^1 N_{i+2}^1)$$

$$+ \frac{1+\mu}{4} cS_{12}(N_{i+1}^2 N_{i+3}^1 + N_{i+1}^1 N_{i+3}^2 + N_i^2 N_{i+2}^1 - 3N_i^1 N_{i+2}^2)$$

$$+ \frac{(1+\mu^2)^{1/2}}{2} cS_{12}[N_i^2(N_{i+1}^1 + N_{i+3}^1) - N_i^1(N_{i+1}^2 + N_{i+3}^2)]$$

$$(2.9)$$

$$\frac{\partial N_i^2}{\partial t} + \mu c\mathbf{e}_i \cdot \nabla_{\mathbf{x}} N_i^2 = cS_{22}(N_{i+1}^2 N_{i+3}^2 - N_i^2 N_{i+2}^2)$$

$$+ \frac{1+\mu}{4} cS_{12}(N_{i+1}^2 N_{i+3}^1 + N_{i+1}^1 N_{i+3}^2 + N_i^1 N_{i+2}^2 - 3N_i^2 N_{i+2}^1)$$

$$+ \frac{(1+\mu^2)^{1/2}}{2} cS_{12}[N_i^1(N_{i+1}^2 + N_{i+3}^2) - N_i^2(N_{i+1}^1 + N_{i+3}^1)]$$

for $i = 1, \ldots, 4$.

The collisional scheme shows that the collisional invariants

$$\Psi = \{\Psi_1^1, \ldots, \Psi_4^1, \Psi_1^2, \ldots, \Psi_4^2\}$$

satisfy twelve relations among which the following

$$\Psi_1^1 + \Psi_3^1 = \Psi_2^1 + \Psi_4^1$$

$$\Psi_1^2 + \Psi_3^2 = \Psi_2^2 + \Psi_4^2$$

$$\Psi_1^1 + \Psi_2^2 = \Psi_2^1 + \Psi_1^2$$

$$\Psi_1^1 + \Psi_4^2 = \Psi_4^1 + \Psi_1^2 ,$$

(2.10)

represent a set of four independent constraints on the components of Ψ. Therefore, the space $\mathcal{M}$ of collisional invariants has dimension equal to four and is spanned by

$$\Phi^{(1)} = \{1, 1, 1, 1, 0, 0, 0, 0\}$$

$$\Phi^{(2)} = \{0, 0, 0, 0, 1, 1, 1, 1\}$$

$$\Phi^{(3)} = \{1, 0, -1, 0, 1, 0, -1, 0\}$$

$$\Phi^{(4)} = \{0, 1, 0, -1, 0, 1, 0, -1\} .$$

(2.11)

These collisional invariants correspond to conservation of mass of the two gas components and of the momentum components of the overall mixture. As for simple gases, if the model allows only one velocity modulus per species, then energy is not an independent macroscopic observable (see Eq.(1.39)). The thermodynamic equilibrium number densities can then be expressed in the usual way as

$$\widehat{\mathbf{N}}^1 = A\{e^{c_3}, e^{c_4}, e^{-c_3}, e^{-c_4}\}$$

$$\widehat{\mathbf{N}}^2 = B\{e^{c_3}, e^{c_4}, e^{-c_3}, e^{-c_4}\} \,,$$

with $A = e^{c_1}$ and $B = e^{c_2}$.

The four Maxwellian parameters A, B, c_3 and c_4 are related to the independent macroscopic observables through

$$\widehat{\nu}_1 = 2A(\cosh c_3 + \cosh c_4)$$

$$\widehat{\nu}_2 = 2B(\cosh c_3 + \cosh c_4)$$

$$\widehat{u}_x = c\frac{(A + \mu B)\sinh c_3}{(A + B)(\cosh c_3 + \cosh c_4)}$$

$$\widehat{u}_y = c\frac{(A + \mu B)\sinh c_4}{(A + B)(\cosh c_3 + \cosh c_4)} \,,$$

with ν_1 and ν_2 defined in (1.25).

2.1.3   *The 6–Velocity Model*

Such a model can be derived following the general procedure given at the beginning of this section. According to the velocity directions visualized in Fig. 2.4 the non-trivial collisions are

$$(\mathbf{v}_1, \mathbf{v}_4) \longleftrightarrow (\mathbf{v}_2, \mathbf{v}_5) \longleftrightarrow (\mathbf{v}_3, \mathbf{v}_6) \quad,$$

with relative velocity $2c$ and probability density $1/3$.

The kinetic equations are

$$\frac{\partial N_1}{\partial t} + c\frac{\partial N_1}{\partial x} = \frac{2}{3}cS(N_2N_5 + N_3N_6 - 2N_1N_4)$$

$$\frac{\partial N_2}{\partial t} + \frac{c}{2}\frac{\partial N_2}{\partial x} + \frac{\sqrt{3}}{2}c\frac{\partial N_2}{\partial y} = \frac{2}{3}cS(N_1N_4 + N_3N_6 - 2N_2N_5)$$

$$\frac{\partial N_3}{\partial t} - \frac{c}{2}\frac{\partial N_3}{\partial x} + \frac{\sqrt{3}}{2}c\frac{\partial N_3}{\partial y} = \frac{2}{3}cS(N_1N_4 + N_2N_5 - 2N_3N_6)$$

$$\frac{\partial N_4}{\partial t} - c\frac{\partial N_4}{\partial x} = \frac{2}{3}cS(N_2N_5 + N_3N_6 - 2N_1N_4) \qquad (2.12)$$

$$\frac{\partial N_5}{\partial t} - \frac{c}{2}\frac{\partial N_5}{\partial x} - \frac{\sqrt{3}}{2}c\frac{\partial N_5}{\partial y} = \frac{2}{3}cS(N_1N_4 + N_3N_6 - 2N_2N_5)$$

$$\frac{\partial N_6}{\partial t} + \frac{c}{2}\frac{\partial N_6}{\partial x} - \frac{\sqrt{3}}{2}c\frac{\partial N_6}{\partial y} = \frac{2}{3}cS(N_1N_4 + N_2N_5 - 2N_3N_6) \ .$$

The collisional invariants satisfy the relations

$$\Psi_1 + \Psi_4 = \Psi_2 + \Psi_5 = \Psi_3 + \Psi_6 \ .$$

Therefore one has six components related by two independent equations, hence $\delta = 4$ and the space of collisional invariants is spanned by the orthogonal basis

$$\Phi^{(1)} = \{1, 1, 1, 1, 1, 1\}$$

$$\Phi^{(2)} = \{1, \frac{1}{2}, -\frac{1}{2}, -1, -\frac{1}{2}, \frac{1}{2}\}$$

$$\Phi^{(3)} = \{0, 1, 1, 0, -1, -1\} \qquad (2.13)$$

$$\Phi^{(4)} = \{1, -1, 1, -1, 1, -1\} \ .$$

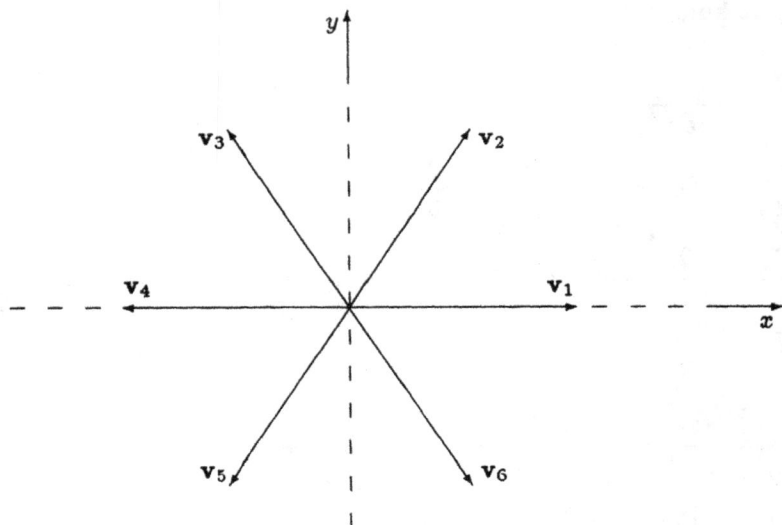

Fig. 2.4 — *Geometry of the planar 6–velocity model*

The first three vectors are related respectively with conservation of mass and momentum components, while the fourth one is related to conservation of "even" and "odd" number densities

$$N_1 + N_3 + N_5 = \zeta$$

$$N_2 + N_4 + N_6 = \nu - \zeta \ ,$$

with $\zeta = \zeta(t, \mathbf{x})$ preserved during the collisions.

This property is a peculiarity of the model that could have been already foreseen by looking carefully at the collisional scheme, but has no precise physical meaning. Then $\Phi^{(4)}$ is a spurious collisional invariant. However, as we shall see in the sections which follow, this feature can be corrected by including triple collisions in the collisional scheme.

2.1.4   *The 6–Velocity Model with Triple Collisions*

This section reports a discrete velocity model with both binary and triple collisions. The model takes into account, in the line of Chapter 1, only symmetric triple collisions.

The 6–velocity discretization already described in Section 2.1.3, generates the following non-trivial collisions

– binary head-on collisions (see Section 2.1.3)

$$(\mathbf{v}_1, \mathbf{v}_4) \longleftrightarrow (\mathbf{v}_2, \mathbf{v}_5) \longleftrightarrow (\mathbf{v}_3, \mathbf{v}_6) \ ;$$

– symmetric triple collisions

$$(\mathbf{v}_1, \mathbf{v}_3, \mathbf{v}_5) \longleftrightarrow (\mathbf{v}_2, \mathbf{v}_4, \mathbf{v}_6) \ .$$

Therefore, the kinetic equations are the ones reported in Eqs.(2.12) with the addition of a triple collision term $J_i^{(3)}$.

The computation of $J_i^{(3)}$ follows the line indicated in Section 1.2, namely a triple collision occurs if, for instance, a particle with velocity $\mathbf{v}_i$ or $\mathbf{v}_{i+2}$ or $\mathbf{v}_{i+4}$, enters the action volume of two particles with velocities $(\mathbf{v}_{i+2}, \mathbf{v}_{i+4})$ or $(\mathbf{v}_i, \mathbf{v}_{i+4})$ or $(\mathbf{v}_i, \mathbf{v}_{i+2})$, respectively during a binary collision. The action volume is geometrically calculated considering that two particles in a binary collision can capture a third particle if their centers are at a distance less than $4\sqrt{\frac{S}{\pi}}$. The three situations

$$\mathbf{v}_i \longleftrightarrow (\mathbf{v}_{i+2}, \mathbf{v}_{i+4}) ,$$

$$\mathbf{v}_{i+2} \longleftrightarrow (\mathbf{v}_i, \mathbf{v}_{i+4}) ,$$

$$\mathbf{v}_{i+4} \longleftrightarrow (\mathbf{v}_i, \mathbf{v}_{i+2}) ,$$

are equally probable.

Detailed calculations [4] yield the following expression of the triple collision term

$$J_i^{(3)} = \frac{5cS^{5/2}}{4\sqrt{\pi}}[N_{i+1}N_{i+3}N_{i+5} - N_iN_{i+2}N_{i+4}] \ , \quad i = 1,\ldots,6 . \quad (2.14)$$

The kinetic equations for the 6–velocity model with both binary and triple collisions are then

$$[\frac{\partial}{\partial t} + c\cos(\frac{i-1}{3}\pi)\frac{\partial}{\partial x} + c\sin(\frac{i-1}{3}\pi)\frac{\partial}{\partial y}]N_i =$$

$$\frac{2}{3}cS(N_{i+1}N_{i+4} + N_{i+2}N_{i+5} - 2N_iN_{i+3})$$

$$+ \frac{5cS^{5/2}}{4\sqrt{\pi}}(N_{i+1}N_{i+3}N_{i+5} - N_iN_{i+2}N_{i+4}) .$$

$$(2.15)$$

The collisional invariants satisfy the three independent relations

$$\Psi_1 + \Psi_4 = \Psi_2 + \Psi_5 = \Psi_3 + \Psi_6$$

$$\Psi_1 + \Psi_3 + \Psi_5 = \Psi_2 + \Psi_4 + \Psi_6 ,$$

so that now $\delta = 3$.

An orthogonal basis for the space $\mathcal{M}$ is given by $\Phi^{(1)}$, $\Phi^{(2)}$, $\Phi^{(3)}$ reported in Eq.(2.13) which correspond, respectively, to conservation of mass and momentum components. Hence the Maxwellians can be written in the form

$$\widehat{N}_1 = Ae^{2c_2} \quad , \quad \widehat{N}_2 = Ae^{c_2+c_3} \quad , \quad \widehat{N}_3 = Ae^{-c_2+c_3} \quad ,$$
$$\widehat{N}_4 = Ae^{-2c_2} \quad , \quad \widehat{N}_5 = Ae^{-c_2-c_3} \quad , \quad \widehat{N}_6 = Ae^{c_2-c_3} \quad ,$$

where $A = e^{c_1}$.

The one-to-one relation between the independent macroscopic observables in equilibrium conditions $\{\widehat{\nu}, \widehat{u}_x, \widehat{u}_y\}$ and $\{A, c_2, c_3\}$ is given by

$$\widehat{\nu} = 2A \left[\cosh 2c_2 + 2\cosh c_2 \cosh c_3\right]$$
$$\widehat{\nu u}_x = 2cA \sinh c_2 \left[2\cosh c_2 + \cosh c_3\right] \qquad (2.16)$$
$$\widehat{\nu u}_y = 2\sqrt{3}cA \sinh c_3 \cosh c_2 \ .$$

The mathematical model defined in Eqs.(2.15) is characterized by two advantages: the triple collision term can simulate some dense gas effects, the number of collisional invariants is the correct one. On the other hand limiting the selection of admissible triple collisions only to the symmetric ones is a strong limitation, although it may be related to a lattice type space discretization. In fact several types of non-symmetric collisions satisfy conservation of momentum and energy (1.14). An example, but not the only one, is

$$\left(\mathbf{v}_i, \mathbf{v}_{i+1}, \mathbf{v}_{i+4}\right) \longleftrightarrow \left(\mathbf{v}_i, \mathbf{v}_{i+2}, \mathbf{v}_{i+5}\right) \ .$$

In this case, however, the application of Eq.(1.15) implies that, even if $a_{ij\ell}^{ghk} = a_{ghk}^{ij\ell}$, as it should be, $A_{ij\ell}^{ghk} \neq A_{ghk}^{ij\ell}$.

Further developments of the discrete Boltzmann equations with a triple collision term need to solve the problem of including in a proper manner all admissible triple collisions.

2.1.5 *The 2×6–Velocity Model with Triple Collisions*

This section describes a model such that the gas particles are allowed to move with two velocity moduli in 6 directions

$$\mathbf{v}_i = c\mathbf{e}_k \quad : \quad i = 2k - 1 \quad , \quad k = 1, \ldots, 6$$

$$\mathbf{v}_i = 2c\mathbf{e}_k \quad : \quad i = 2k \quad \ , \quad k = 1, \ldots, 6 \ ,$$

where

$$\mathbf{e}_k = \cos\left[(k-1)\frac{\pi}{3}\right]\mathbf{i} + \sin\left[(k-1)\frac{\pi}{3}\right]\mathbf{j} \ ,$$

and where the ratio of the two moduli is properly chosen in order to have admissible collisions between particles with different moduli.

The collisional scheme is the following

1) Binary head-on collisions

$$(\mathbf{v}_i, \mathbf{v}_{i+6}) \longleftrightarrow (\mathbf{v}_{i+2}, \mathbf{v}_{i+8}) \longleftrightarrow (\mathbf{v}_{i+4}, \mathbf{v}_{i+10}) \ , \quad i = 1, 2$$

with probability density 1/3 and relative velocity $2c$ if i is odd and $4c$ if i is even.

2) Binary collisions at angle

$$\{\mathbf{v}_i, \mathbf{v}_{i+5}\} \longleftrightarrow \{\mathbf{v}_{i+6}, \mathbf{v}_{i+3}\} \quad , \quad i = 2k - 1 \quad , \quad k = 1, \ldots, 6$$

with probability density $1/2$ and relative velocity $\sqrt{7}c$.

3) Symmetric triple collisions

$$\{\mathbf{v}_i, \mathbf{v}_{i+4}, \mathbf{v}_{i+8}\} \longleftrightarrow \{\mathbf{v}_{i+2}, \mathbf{v}_{i+6}, \mathbf{v}_{i+10}\} \quad , \quad i = 1, 2$$

with probability density $1/2$ and relative velocity $\sqrt{3}c$ if i is odd and $2\sqrt{3}c$ if i is even.

The kinetic equations then are

$$
[\frac{\partial}{\partial t} + c\cos(\frac{i-1}{6}\pi)\frac{\partial}{\partial x} + c\sin(\frac{i-1}{6}\pi)\frac{\partial}{\partial y}]N_i
$$

$$
= \frac{2}{3}cS(N_{i+2}N_{i+8} + N_{i+4}N_{i+10} - 2N_iN_{i+6})
$$

$$
+ \frac{\sqrt{7}}{2}cS[N_{i+6}(N_{i+3} + N_{i+11}) - N_i(N_{i+5} + N_{i+9})]
$$

$$
+ \frac{5}{4\sqrt{\pi}}cS^{5/2}(N_{i+2}N_{i+6}N_{i+10} - N_iN_{i+4}N_{i+8})
$$

(2.17)

if i is odd, and

$$
[\frac{\partial}{\partial t} + 2c\cos(\frac{i-2}{6}\pi)\frac{\partial}{\partial x} + 2c\sin(\frac{i-2}{6}\pi)\frac{\partial}{\partial y}]N_i
$$

$$
= \frac{4}{3}cS(N_{i+2}N_{i+8} + N_{i+4}N_{i+10} - 2N_iN_{i+6})
$$

$$
+ \frac{\sqrt{7}}{2}cS[N_{i+10}N_{i+1} + N_{i+2}N_{i+9} - N_i(N_{i+3} + N_{i+7})]
$$

$$
+ \frac{5}{2\sqrt{\pi}}cS^{5/2}(N_{i+2}N_{i+6}N_{i+10} - N_iN_{i+4}N_{i+8})
$$

(2.18)

if i is even.

The collision invariants satisfy the constraints

$$\Psi_i + \Psi_{i+6} = \Psi_{i+2} + \Psi_{i+8} = \Psi_{i+4} + \Psi_{i+10} \quad , \quad i = 1, 2$$

$$\Psi_i + \Psi_{i+5} = \Psi_{i+3} + \Psi_{i+6} \quad , \quad i = 2k - 1 \quad , \quad k = 1, \ldots, 6$$

$$\Psi_i + \Psi_{i+4} + \Psi_{i+8} = \Psi_{i+2} + \Psi_{i+6} + \Psi_{i+10} \quad , \quad i = 1, 2$$

which result to form a set of eight independent quantities; thus dim $\mathcal{M} = 12 - 8 = 4$, according to the conservation of mass, momentum components and energy, which is now an independent quantity thanks to the two velocity moduli allowed by the model.

A suitable orthogonal basis spanning the linear subspace $\mathcal{M} \subset \mathbf{R}^{12}$ is given by the following set of collision invariants

$$\Phi^{(\chi)} = \{\Phi_i^{(\chi)}\} \in \mathbf{R}^{12} \quad , \quad \chi = 1, \ldots, 4 \, ,$$

where

$$\Phi_i^{(1)} = \begin{cases} 1 & \text{if } i \text{ is odd} \\ 0 & \text{if } i \text{ is even} \end{cases}$$

$$\Phi_i^{(2)} = \begin{cases} 0 & \text{if } i \text{ is odd} \\ 1 & \text{if } i \text{ is even} \end{cases}$$

$$\Phi_i^{(3)} = \frac{\mathbf{v}_i \cdot \mathbf{i}}{c}$$

$$\Phi_i^{(4)} = \frac{\mathbf{v}_i \cdot \mathbf{j}}{c} \, .$$

The Maxwellians are then given by

$$\widehat{N}_i = \begin{cases} A \exp\left[c_x \frac{v_{ix}}{c} + c_y \frac{v_{iy}}{c}\right] & \text{if } i \text{ is odd} \\[2mm] B \exp\left[c_x \frac{v_{ix}}{c} + c_y \frac{v_{iy}}{c}\right] & \text{if } i \text{ is even}, \end{cases} \tag{2.19}$$

where $A = e^{c_\nu}$ and $B = e^{c_T}$.

Moreover the hydrodynamic moments W_χ, $\chi = 1, \ldots, 4$ are related with the macroscopic observables by

$$\begin{aligned} \nu &= W_1 + W_2 \\[2mm] u_x &= \frac{cW_3}{W_1 + W_2} \\[2mm] u_y &= \frac{cW_4}{W_1 + W_2} \\[2mm] T &= \frac{c^2}{3\mathcal{R}}\left[\frac{W_1 + 4W_2}{W_1 + W_2} - \frac{W_3^2 + W_4^2}{(W_1 + W_2)^2}\right] \end{aligned} \tag{2.20}$$

where $\mathcal{R} = k_B/m$ and, in the Maxwellian state, with the equilibrium parameters A, B, c_x, c_y by

$$\widehat{W}_1 = \frac{\widehat{\nu}}{3c^2}(4c^2 - \widehat{u}_x^2 - \widehat{u}_y^2 - 3\mathcal{R}\widehat{T}) = 2A\left(\cosh c_x + 2\cosh\frac{c_x}{2}\cosh\frac{\sqrt{3}}{2}c_y\right)$$

$$\widehat{W}_2 = \frac{\widehat{\nu}}{3c^2}(3\mathcal{R}\widehat{T} - c^2 + \widehat{u}_x^2 + \widehat{u}_y^2) = 2B\left(\cosh 2c_x + 2\cosh c_x \cosh\sqrt{3}c_y\right)$$

$$\widehat{W}_3 = \frac{\widehat{\nu}\widehat{u}_x}{c} = 2A\left(\sinh c_x + \sinh\frac{c_x}{2}\cosh\frac{\sqrt{3}}{2}c_y\right)$$
$$\qquad\qquad + 4B\left(\sinh 2c_x + \sinh c_x \cosh\sqrt{3}c_y\right)$$

$$\widehat{W}_4 = \frac{\widehat{\nu}\widehat{u}_y}{c} = 2\sqrt{3}A\cosh\frac{c_x}{2}\sinh\frac{\sqrt{3}}{2}c_y + 4\sqrt{3}B\cosh c_x \sinh\sqrt{3}c_y .$$

$$\tag{2.21}$$

Also this model, as the one of the preceding section, is subject to the limitation, discussed in 2.1.4, to triple collision of symmetric-type.

2.2 Spatial Models

Broadwell proposed in two papers [5,6] a discrete velocity model with six admissible velocities, which has been used for several applications by various authors. This model and its generalization to a binary gas mixtures are presented here. Later on Cabannes [7] proposed a model with fourteen velocities obtained joining Brodwell's model to a spatial 8–velocity model [1]. In Section 2.2.3 the generalization of this model to a gas mixture is presented. Finally the last section deals shortly with a set of models with a large number of velocities proposed by Shizuta and Kawashima [9–11]

2.2.1 *The Broadwell Model for a One-Component Gas*

Consider a monoatomic gas in the physical $\mathbf{R}^3$ space and let the velocities have the same modulus and directions parallel to the axes x , y and z, as shown in Fig. 2.5.

The allowed velocities can be written as

$$i = 1, \ldots, 6 : \qquad \mathbf{v}_i = c\mathbf{e}_i , \qquad (2.22)$$

with

$$\mathbf{e}_1 = \mathbf{i} , \ \ \mathbf{e}_2 = \mathbf{j} , \ \ \mathbf{e}_3 = \mathbf{k} , \ \ \mathbf{e}_4 = -\mathbf{i} , \ \ \mathbf{e}_5 = -\mathbf{j} , \ \ \mathbf{e}_6 = -\mathbf{k} . \ (2.23)$$

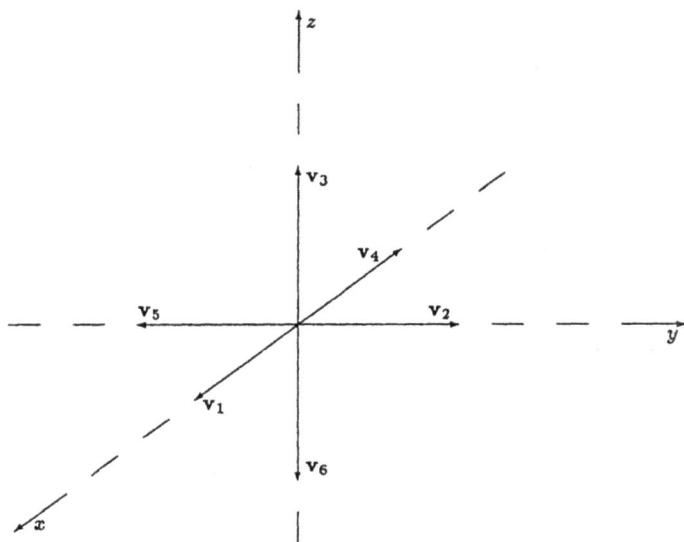

Fig. 2.5 — *Geometry of the Broadwell model*

The admissible collisions are only of head-on type

$$(\mathbf{v}_1, \mathbf{v}_4) \longleftrightarrow (\mathbf{v}_2, \mathbf{v}_5) \longleftrightarrow (\mathbf{v}_3, \mathbf{v}_6)$$

with probability density

$$a_{i,i+3}^{j,j+3} = \frac{1}{3} \quad , \quad i \neq j = 1, 2, 3$$

and relative velocity $2c$ and the index i is to be intended modulo 6.

The kinetic equations are

$$\frac{\partial N_i}{\partial t} + c\mathbf{e}_i \cdot \nabla_{\mathbf{x}} N_i = \frac{2}{3} cS(N_{i+1}N_{i+4} + N_{i+2}N_{i+5} - 2N_iN_{i+3}) \ , \quad (2.24)$$

for $i = 1, \ldots, 6$.

The collisional invariants satisfy the relations

$$\Psi_1 + \Psi_4 = \Psi_2 + \Psi_5 = \Psi_3 + \Psi_6 , \qquad (2.25)$$

then one has six arbitrary quantities and two independent equations. Thus, the space $\mathcal{M}$ has dimension equal to four and is spanned by the orthogonal basis

$$\Phi^{(1)} = \{1, 1, 1, 1, 1, 1\}$$
$$\Phi^{(2)} = \{1, 0, 0, -1, 0, 0\}$$
$$\Phi^{(3)} = \{0, 1, 0, 0, -1, 0\} \qquad (2.26)$$
$$\Phi^{(4)} = \{0, 0, 1, 0, 0, -1\} .$$

These collisional invariants correspond, respectively, to conservation of mass and momentum.

The Maxwellian number densities are given by

$$\widehat{N}_i = A e^{c_{i+1}} \quad , \quad \widehat{N}_{i+3} = A e^{-c_{i+1}} \qquad\qquad i = 1, 2, 3 \qquad (2.27)$$

where $A = e^{c_1}$.

The four Maxwellian parameters A, c_2, c_3 and c_4 are one-to-one related to $\widehat{\nu}, \widehat{u}_x, \widehat{u}_y, \widehat{u}_z$ by

$$\widehat{\nu} = 2A(\cosh c_2 + \cosh c_3 + \cosh c_4)$$
$$\widehat{\nu}\widehat{\mathbf{u}} = 2Ac\{\sinh c_2, \sinh c_3, \sinh c_4\} .$$

In several fluid-dynamic applications, the Broadwell model is used in its one-dimensional form. In this case the kinetic equations are

$$\frac{\partial N_1}{\partial t} + c\frac{\partial N_1}{\partial x} = \frac{4}{3}\, cS(N_2^2 - N_1 N_4)$$

$$\frac{\partial N_2}{\partial t} = \frac{2}{3}\, cS(N_1 N_4 - N_2^2) \tag{2.28}$$

$$\frac{\partial N_4}{\partial t} - c\frac{\partial N_4}{\partial x} = \frac{4}{3}\, cS(N_2^2 - N_1 N_4)\ ,$$

with $N_2 = N_3 = N_5 = N_6$.

The Maxwellian number densities will then be given by

$$\widehat{\mathbf{N}} = \{\widehat{N}_1, \widehat{N}_2, \widehat{N}_4\} = A\{e^b, 1, e^{-b}\} \tag{2.29}$$

substituting $c_2 = b$, $c_3 = c_4 = 0$ in (2.27). The hydrodynamic moments $\widehat{W}_1$ and $\widehat{W}_2$ in equilibrium conditions are expressed in terms of the equilibrium parameters A and b by

$$\widehat{W}_1 = \widehat{\nu} = \widehat{N}_1 + 4\widehat{N}_2 + \widehat{N}_4 = 2A(\cosh b + 2)$$

$$\widehat{W}_2 = \widehat{\nu}\widehat{u}_x = c(\widehat{N}_1 - \widehat{N}_4) = 2Ac\sinh b\ \ . \tag{2.30}$$

The one-to-one map (2.30) can be inverted so that the Euler equations are

$$\frac{\partial \widehat{\nu}}{\partial t} + \frac{\partial}{\partial x}(\widehat{\nu}\widehat{u}_x) = 0$$

$$\frac{\partial}{\partial t}(\widehat{\nu}\widehat{u}_x) + \frac{\partial}{\partial x}\left[\frac{\widehat{\nu}c}{3}(-c + \sqrt{c^2 + 3\widehat{u}_x^2})\right] = 0\ . \tag{2.31}$$

2.2.2 Broadwell Model for Binary Gas Mixtures

The Broadwell model is also of great interest because of its applications to gas mixtures. Here the special case of two gas components is considered; the derivation of the model for a mixture with an arbitrary number of components is then only a matter of generalization and can be done following the procedure explained in Section 1.2.2.

Let the two gas components have molecular masses m_1 and m_2, respectively. As usual we will select the following velocities

$$i = 1, \ldots, 6 : \qquad \mathbf{v}_i^1 = c\mathbf{e}_i \quad , \quad \mathbf{v}_i^2 = \mu c\mathbf{e}_i$$

where $\mu = m_1/m_2 < 1$ and the unit vectors $\mathbf{e}_i$ are defined in Eq.(2.23). The collisional scheme is

a) Head-on collisions between molecules of the same gas with probability density equal to $1/3$

   a_1) $(\mathbf{v}_i^1, \mathbf{v}_{i+3}^1) \longleftrightarrow (\mathbf{v}_j^1, \mathbf{v}_{j+3}^1)$, $j \neq i = 1, 2, 3$
   with relative velocity $2c$;

   a_2) $(\mathbf{v}_i^2, \mathbf{v}_{i+3}^2) \longleftrightarrow (\mathbf{v}_j^2, \mathbf{v}_{j+3}^2)$, $j \neq i = 1, 2, 3$
   with relative velocity $2\mu c$;

b) Head-on collisions between molecules of different gases

$$(\mathbf{v}_i^1, \mathbf{v}_{i+3}^2) \longleftrightarrow (\mathbf{v}_j^1, \mathbf{v}_{j+3}^2) \quad , \quad j \neq i = 1, \ldots, 6$$

with probability density equal to $1/6$ and relative velocity $c(1+\mu)$;

c) **Exchange collisions at angle between particles of the two gases**

$$(\mathbf{v}_i^1, \mathbf{v}_j^2) \longleftrightarrow (\mathbf{v}_j^1, \mathbf{v}_i^2) \quad , \quad i, j = 1, \ldots, 6 \quad , \quad j \neq i, i+3$$

with probability density $1/2$ and relative velocity $c(1 + \mu^2)^{1/2}$.

The kinetic equations are, then

$$\frac{\partial N_i^1}{\partial t} + c e_i \cdot \nabla_{\mathbf{x}} N_i^1 = \frac{2}{3} c S_{11} (N_{i+1}^1 N_{i+4}^1 + N_{i+2}^1 N_{i+5}^1 - 2 N_i^1 N_{i+3}^1)$$

$$+ \frac{1+\mu}{6} c S_{12} (N_{i+1}^1 N_{i+4}^2 + N_{i+2}^1 N_{i+5}^2 + N_{i+3}^1 N_i^2$$

$$+ N_{i+4}^1 N_{i+1}^2 + N_{i+5}^1 N_{i+2}^2 - 5 N_i^1 N_{i+3}^2)$$

$$+ \frac{\sqrt{1+\mu^2}}{2} c S_{12} [N_i^2 (N_{i+1}^1 + N_{i+2}^1 + N_{i+4}^1 + N_{i+5}^1)$$

$$- N_i^1 (N_{i+1}^2 + N_{i+2}^2 + N_{i+4}^2 + N_{i+5}^2)]$$

$$(2.32)$$

$$\frac{\partial N_i^2}{\partial t} + \mu c e_i \cdot \nabla_{\mathbf{x}} N_i^2 = \frac{2}{3} \mu c S_{22} (N_{i+1}^2 N_{i+4}^2 + N_{i+2}^2 N_{i+5}^2 - 2 N_i^2 N_{i+3}^2)$$

$$+ \frac{1+\mu}{6} c S_{12} (N_{i+1}^2 N_{i+4}^1 + N_{i+2}^2 N_{i+5}^1 + N_{i+3}^2 N_i^1$$

$$+ N_{i+4}^2 N_{i+1}^1 + N_{i+5}^2 N_{i+2}^1 - 5 N_i^2 N_{i+3}^1)$$

$$+ \frac{\sqrt{1+\mu^2}}{2} c S_{12} [N_i^1 (N_{i+1}^2 + N_{i+2}^2 + N_{i+4}^2 + N_{i+5}^2)$$

$$- N_i^2 (N_{i+1}^1 + N_{i+2}^1 + N_{i+4}^1 + N_{i+5}^1)]$$

for $i = 1, \ldots, 6$.

The dimension δ of the collisional invariants space $\mathcal{M}$, for this model, is equal to five, according to the fact that the masses of the two gases and the three momentum components of the whole mixture are preserved.

The orthogonal basis of $\mathcal{M}$ can be written generalizing the result for a one component gas to a binary mixture

$$\Phi^{(1)} = \{1, 1, 1, 1, 1, 1, 0, 0, 0, 0, 0, 0\}$$

$$\Phi^{(2)} = \{0, 0, 0, 0, 0, 0, 1, 1, 1, 1, 1, 1\}$$

$$\Phi^{(3)} = \{1, 0, 0, -1, 0, 0, 1, 0, 0, -1, 0, 0\}$$

$$\Phi^{(4)} = \{0, 1, 0, 0, -1, 0, 0, 1, 0, 0, -1, 0\}$$

$$\Phi^{(5)} = \{0, 0, 1, 0, 0, -1, 0, 0, 1, 0, 0, -1\}$$

and consequently the Maxwellians are

$$\widehat{\mathbf{N}}^1 = A\{e^{c_3}, e^{c_4}, e^{c_5}, e^{-c_3}, e^{-c_4}, e^{-c_5}\}$$
$$\widehat{\mathbf{N}}^2 = B\{e^{c_3}, e^{c_4}, e^{c_5}, e^{-c_3}, e^{-c_4}, e^{-c_5}\} \tag{2.33}$$

where $A = e^{c_1}$ and $B = e^{c_2}$. The five Maxwellian parameters A, B, c_3, c_4

and c_5 are one-to-one related to $\widehat{\nu}_1, \widehat{\nu}_2, \widehat{u}_x, \widehat{u}_y, \widehat{u}_z$ by

$$\widehat{\nu}_1 = 2A(\cosh c_3 + \cosh c_4 + \cosh c_5)$$

$$\widehat{\nu}_2 = 2B(\cosh c_3 + \cosh c_4 + \cosh c_5)$$

$$\widehat{u}_x = \frac{c(A + \mu B)\sinh c_3}{(A + B)(\cosh c_3 + \cosh c_4 + \cosh c_5)}$$

$$\widehat{u}_y = \frac{c(A + \mu B)\sinh c_4}{(A + B)(\cosh c_3 + \cosh c_4 + \cosh c_5)}$$

$$\widehat{u}_z = \frac{c(A + \mu B)\sinh c_5}{(A + B)(\cosh c_3 + \cosh c_4 + \cosh c_5)}\ .$$

2.2.3   Cabannes-Type Models

This section deals with a spatial model with two velocity moduli per gas component. The original model has been proposed by Cabannes [7] who considered the case of a single monoatomic gas. The model has then been generalized to gas mixtures with an arbitrary number of gas components [8].

The velocity discretization for a one-component gas, is the following

- six velocities are directed from the center **C** of a cube to the centers of its six faces; the modulus of these velocities is c;

- eight velocities are directed from the center **C** of the cube to its eight vertices; the modulus of these velocities is $\sqrt{3}c$.

In order to deal with the 14–velocity model for a multi-component gas mixture, it is useful to refer the vertices of the cube in Fig. 2.6 to triplets of numbers relative to the faces forming the vertex

$$
\begin{aligned}
\text{vertex 1} &\longleftrightarrow (4\ 2\ 3) \\
\text{vertex 2} &\longleftrightarrow (1\ 2\ 3) \\
\text{vertex 3} &\longleftrightarrow (4\ 5\ 3) \\
\text{vertex 4} &\longleftrightarrow (1\ 5\ 3) \\
\text{vertex 5} &\longleftrightarrow (4\ 2\ 6) \\
\text{vertex 6} &\longleftrightarrow (1\ 2\ 6) \\
\text{vertex 7} &\longleftrightarrow (4\ 5\ 6) \\
\text{vertex 8} &\longleftrightarrow (1\ 5\ 6)\ .
\end{aligned}
\tag{2.34}
$$

Then, the velocity discretization for the lightest gas can be defined as follows

$$
\begin{aligned}
\mathbf{v}_1^1 &= -\mathbf{v}_4^1 &&= c\{1,0,0\} \\
\mathbf{v}_2^1 &= -\mathbf{v}_5^1 &&= c\{0,1,0\} \\
\mathbf{v}_3^1 &= -\mathbf{v}_6^1 &&= c\{0,0,1\} \\
\mathbf{V}_{123}^1 &= -\mathbf{V}_{456}^1 &&= c\{1,1,1\} \\
\mathbf{V}_{153}^1 &= -\mathbf{V}_{426}^1 &&= c\{1,-1,1\} \\
\mathbf{V}_{126}^1 &= -\mathbf{V}_{453}^1 &&= c\{1,1,-1\} \\
\mathbf{V}_{156}^1 &= -\mathbf{V}_{423}^1 &&= c\{1,-1,-1\}\ ,
\end{aligned}
$$

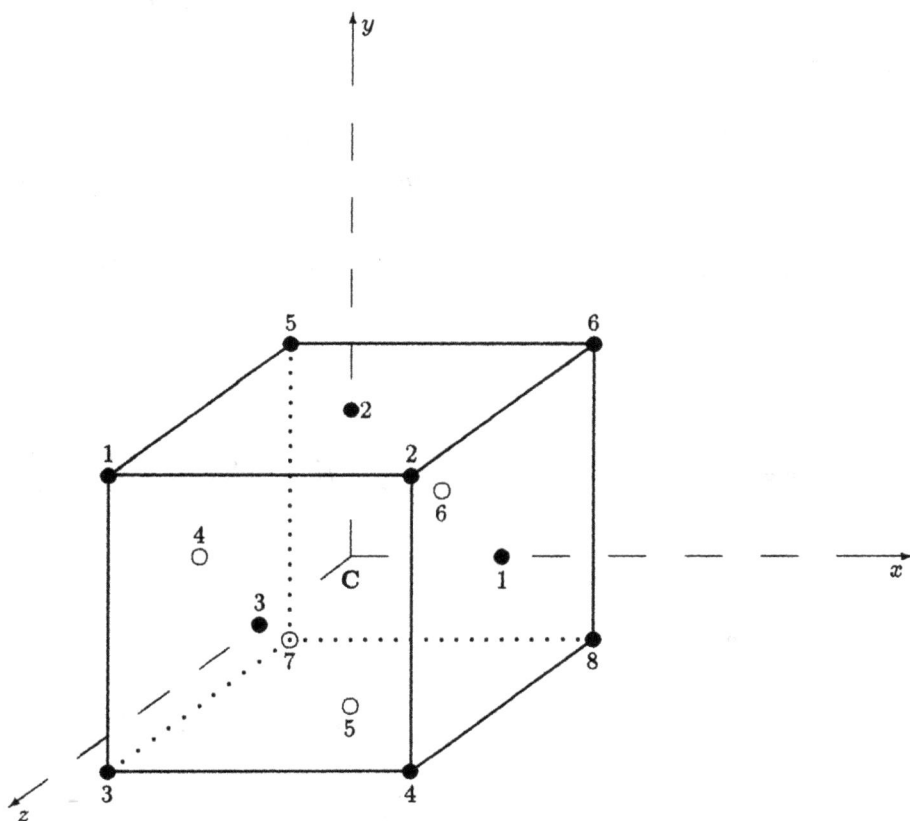

Fig. 2.6 — *Geometry of the 14–velocity model*

and, for the mixture, by

$$i = 1, \ldots, 6 \; : \quad \mathbf{v}_i^p = \mu_p \mathbf{v}_i^1 \quad , \quad N_i^p = N_i^p(t, x, y, z)$$

$$(jk\ell) = 1, \ldots, 8 \; : \quad \mathbf{V}_{jk\ell}^p = \mu_p \mathbf{V}_{jk\ell}^1 \quad , \quad M_{jk\ell}^p = M_{jk\ell}^p(t, x, y, z) \,,$$

where the allowed triplets $(jk\ell)$ are the one indicated in Eq.(2.34), $\mu_p = m_1/m_p \leq 1$, $p = 1, \ldots, P$ and m_p is the molecular mass of the p-species.

The admissible collisions for such a model are the following with $\check{j} = j + 3$, $\check{k} = k + 3$, $\check{\ell} = \ell + 3$

a) **Head-on collisions**

$$\forall p, r = 1, \ldots, P \;, \; \forall j, k = 1, \ldots, 6 \;, \; j \neq k \; : \; (\mathbf{v}_{\check{j}}^p, \mathbf{v}_{\check{j}}^r) \longleftrightarrow (\mathbf{v}_k^p, \mathbf{v}_k^r)$$

with transition rate $\frac{1}{6}(\mu_p + \mu_r)cS_{pr}$;

b) **Exchange collisions**

$$\forall p \neq r = 1, \ldots, P \;, \; \forall k \neq j, \check{j} \; : \; (\mathbf{v}_{\check{j}}^p, \mathbf{v}_k^r) \longleftrightarrow (\mathbf{v}_k^p, \mathbf{v}_{\check{j}}^r)$$

with transition rate $\frac{1}{2}(\mu_p^2 + \mu_r^2)^{1/2}cS_{pr}$;

c) **Collisions at angle**

$$\forall p, r = 1, \ldots, P \;, \; \forall k = j + 1, \check{j} + 1 \;, \; \forall \ell = j + 2, \check{j} + 2 \; :$$

$$(\mathbf{v}_{\check{j}}^p, \mathbf{V}_{\check{j}k\ell}^r) \longleftrightarrow (\mathbf{v}_j^p, \mathbf{V}_{jk\ell}^r)$$

with transition rate $\frac{1}{2}\left[(\mu_p + \mu_r)^2 + 2\mu_r^2\right]^{1/2} cS_{pr}$;

d) Head-on collisions

$$\forall p, r = 1, \ldots, P \ , \ \forall(jk\ell) \ , \ \forall(ghi) \ :$$

$$(\mathbf{V}_{jk\ell}^p, \mathbf{V}_{jk\bar{\ell}}^r) \longleftrightarrow (\mathbf{V}_{ghi}^p, \mathbf{V}_{\bar{g}hi}^r)$$

with transition rate $\frac{\sqrt{3}}{8}(\mu_p + \mu_r)cS_{pr}$;

e) Collisions at angle

$$\forall p, r = 1, \ldots, P \ , \ \forall(jk\ell) \ :$$

$$(\mathbf{V}_{jk\ell}^p, \mathbf{V}_{jk\bar{\ell}}^r) \longleftrightarrow (\mathbf{V}_{jk\bar{\ell}}^p, \mathbf{V}_{jk\ell}^r) \longleftrightarrow (\mathbf{V}_{jk\bar{\ell}}^p, \mathbf{V}_{jk\ell}^r) \longleftrightarrow (\mathbf{V}_{jk\ell}^p, \mathbf{V}_{jk\bar{\ell}}^r)$$

with transition rate $\frac{1}{4}[(\mu_p - \mu_r)^2 + 2(\mu_p + \mu_r)^2]^{1/2}cS_{pr}$;

f) Exchange collisions $\forall p \neq r = 1, \ldots, P \ , \ \forall(ghi) = (jk\breve{\ell}), (j\breve{k}\ell), (\breve{j}k\ell)$:

$$(\mathbf{V}_{jk\ell}^p, \mathbf{V}_{ghi}^r) \longleftrightarrow (\mathbf{V}_{ghi}^p, \mathbf{V}_{jk\ell}^r)$$

with transition rate $\frac{1}{2}[(\mu_p + \mu_r)^2 + 2(\mu_p - \mu_r)^2]^{1/2}cS_{pr}$.

Accordingly, one can write in a compact form the $14 \times P$ kinetic equations

$$
\frac{\partial}{\partial t} N_j^p + \mathbf{v}_j^p \cdot \nabla_{\mathbf{x}} N_j^p = \frac{c}{2} \sum_{r=1}^{P} S_{pr} \left\{ \left[\frac{1}{3}(\mu_p + \mu_r)(\sum_{k \neq j} N_k^p N_k^r - 5 N_j^p N_j^r) \right] \right.
$$

$$
+ \sqrt{\mu_p^2 + \mu_r^2} \sum_{k=j+1,j+2,j+1,j+2} (N_k^p N_j^r - N_j^p N_k^r)
$$

$$
\left. + \sqrt{(\mu_p + \mu_r)^2 + 2\mu_r^2} \sum_{\substack{k=j+1,j+1 \\ l=j+2,j+2}} (N_j^p M_{jkl}^r - N_j^p M_{jkl}^r) \right\}
$$

$$(2.35a)$$

$$
\frac{\partial}{\partial t} M_{jkl}^p + \mathbf{V}_{jkl}^p \cdot \nabla_{\mathbf{x}} M_{jkl}^p
$$

$$
= \frac{c}{2} \sum_{r=1}^{P} S_{pr} \left\{ \sqrt{(\mu_p + \mu_r)^2 + 2\mu_p^2} \sum_{(ghi)} (M_{ghi}^p N_g^r - M_{jkl}^p N_j^r) \right.
$$

$$
+ \frac{\sqrt{3}}{4}(\mu_p + \mu_r) \sum_{\substack{g=1,4 \\ h=g+1,g+1 \\ i=g+2,g+2}} (M_{ghi}^p M_{ghi}^r - M_{jkl}^p M_{jkl}^r)
$$

$$
+ \sqrt{(\mu_p + \mu_r)^2 + 2(\mu_p - \mu_r)^2} \sum_{(ghi)=(jkl),(jkl),(jkl)} (M_{ghi}^p M_{jkl}^r - M_{jkl}^p M_{ghi}^r)
$$

$$
+ \frac{1}{2}\sqrt{(\mu_p - \mu_r)^2 + 2(\mu_p + \mu_r)^2} [(M_{jkl}^p + M_{jkl}^p + M_{jkl}^p)M_{jkl}^r
$$

$$
+ (M_{jkl}^p + M_{jkl}^p)M_{jkl}^r + (M_{jkl}^p + M_{jkl}^p)M_{jkl}^r + (M_{jkl}^p + M_{jkl}^p)M_{jkl}^r
$$

$$
\left. - 3M_{jkl}^p (M_{jkl}^r + M_{jkl}^r + M_{jkl}^r)] \right\}.
$$

$$(2.35b)$$

Setting $P = 1$ in Eqs.(2.35), the well-known [7] Cabannes 14–velocity model is obtained. Setting $M_{jk\ell} = 0$ in (2.35a) gives the kinetic equations of the Broadwell model for a mixture with P components.

The dimension of the space $\mathcal{M}$ can be found as described in Section 1.4.2. Since $\tilde{\delta} = 5$ (see ref.[7]) and the number of different moduli $\tilde{n}$ is equal to 2, the result is

$$\delta = 5 + 2(P - 1) = 3 + 2P ,$$

according to the fact that the model preserves the P masses, the P energies and the overall momentum.

The orthogonal basis of the space of collisional invariants can be obtained by a suitable generalization of Eq.(1.42). In fact, the basis vectors

$$\mathbf{\Psi}^{(\chi)} = \{\Psi_1^{1(\chi)}, \ldots, \Psi_{14}^{1(\chi)}, \ldots, \Psi_1^{P(\chi)}, \ldots, \Psi_{14}^{P(\chi)}\} \in \mathbb{R}^{14P} ,$$

for $\chi = 1, \ldots, 2P + 3$, are

$$\chi = 1, \ldots, P , \quad \Psi_i^{p(\chi)} = \begin{cases} 1 & \text{if } p = \chi \\ 0 & \text{otherwise} \end{cases}$$

$$\chi = P + 1, \ldots, 2P , \quad \Psi_i^{p(\chi)} = \begin{cases} \mu_p & \text{if } p = \chi \text{ and } i = 1, \ldots, 6 \\ 3\mu_p & \text{if } p = \chi \text{ and } i = 7, \ldots, 14 \\ 0 & \text{otherwise} \end{cases}$$

$$\Psi_i^{p(2P+1)} = X_i$$

$$\Psi_i^{p(2P+2)} = Y_i$$

$$\Psi_i^{p(2P+3)} = Z_i \, ,$$

where X_i, Y_i and Z_i are the i–th component of the vectors

$$\mathbf{X} = \{1, 0, 0, -1, 0, 0, -1, 1, -1, 1, -1, 1, -1, 1\}$$

$$\mathbf{Y} = \{0, 1, 0, 0, -1, 0, 1, 1, -1, -1, 1, 1, -1, -1\}$$

$$\mathbf{Z} = \{0, 0, 1, 0, 0, -1, 1, 1, 1, 1, -1, -1, -1, -1\} \, .$$

The first P vectors correspond to conservation of mass per species of the P–component gas mixture, the second set of P vectors corresponds to conservation of energy per species and the last three vectors correspond to conservation of the overall momentum.

The fact that energy of each gas component is a collisional invariant assures necessarily conservation of energy of the overall mixture. On the other hand such situation seems to be not very satisfactory on a physical ground, because the model does not allow exchanges of energy between the different species of gas; as a matter of fact this is a direct consequence of assuming the same momentum discretization for the whole mixture. This point has been discussed in ref.[8], where the reader may be addressed for further details.

After this, one can derive the Maxwellian number densities

$$\widehat{\mathbf{N}}^p = A_p\, B_p\, \{e^{c_x}, e^{c_y}, e^{c_z}, e^{-c_x}, e^{-c_y}, e^{-c_z}\}$$

$$\widehat{\mathbf{M}}^p = A_p(B_p)^3\{e^{-c_x+c_y+c_z}, e^{c_x+c_y+c_z}, e^{-c_x-c_y+c_z}, e^{c_x-c_y+c_z},$$

$$e^{-c_x+c_y-c_z}, e^{c_x+c_y-c_z}, e^{-c_x-c_y-c_z}, e^{c_x-c_y-c_z}\}\,,$$

$$(2.36)$$

$\forall i = 1, \ldots, 6\,,\ \forall(jk\ell) = 1, \ldots, 8\,,\ \forall p = 1, \ldots, P\,,\ $ where

$$A_p = \exp[c_{\nu_p}]\quad,\quad B_p = \exp[\mu_p c_{T_p}]\,,$$

$\widehat{\mathbf{N}}^p = \{\widehat{N}_1^p, \ldots, \widehat{N}_6^p\} \in \mathbb{R}^6$ and $\widehat{\mathbf{M}}^p \in \mathbb{R}^8$ is the equivalent definition of the Maxwellian number densities $\widehat{M}_{jk\ell}^p$.

The $2P+3$ Maxwellian parameters $c_x, c_y, c_z, A_p, B_p\,,\ p = 1, \ldots, P$ are related to the P densities, the P energies and to the three components of the overall momentum of the gas mixture through

$$\widehat{\nu}_p = 2A_pB_p[C_p + 4(B_p)^2 D_p]$$

$$\widehat{\nu u}_x = 2c\sinh c_x \sum_{p=1}^{P} \mu_p A_p B_p[1 + 4(B_p)^2 \cosh c_y \cosh c_z]$$

$$\widehat{\nu u}_y = 2c\sinh c_y \sum_{p=1}^{P} \mu_p A_p B_p[1 + 4(B_p)^2 \cosh c_x \cosh c_z]$$

$$(2.37)$$

$$\widehat{\nu u}_z = 2c\sinh c_z \sum_{p=1}^{P} \mu_p A_p B_p[1 + 4(B_p)^2 \cosh c_x \cosh c_y]$$

$$\widehat{T}_p = \frac{\mu_p^2 m_p c^2}{3k_B}\left\{\frac{C_p + 12(B_p)^2 D_p}{C_p + 4(B_p)^2 D_p} - \frac{E_p}{[C_p + 4(B_p)^2 D_p]^2}\right\}\,.$$

where

$$C_p = \cosh c_x + \cosh c_y + \cosh c_z$$

$$D_p = \cosh c_x \cosh c_y \cosh c_z$$

$$E_p = \sinh^2 c_x [1 + 4(B_p)^2 \cosh c_y \cosh c_z]^2$$
$$+ \sinh^2 c_y [1 + 4(B_p)^2 \cosh c_x \cosh c_z]^2$$
$$+ \sinh^2 c_z [1 + 4(B_p)^2 \cosh c_x \cosh c_y]^2 .$$

In particular, the drift velocity vanishes if and only if $c_x = c_y = c_z = 0$. In this case $C_p = 3$, $D_p = 1$ and $E_p = 0$, and hence

$$\widehat{\nu}_p = 6A_p B_p + 8A_p(B_p)^3$$

$$\widehat{T}_p = \frac{\mu_p^2 m_p c^2}{k_B} \frac{1 + 4(B_p)^2}{3 + 4(B_p)^2}$$

can be inverted giving

$$\widehat{N}_i^p = A_p B_p = \frac{\widehat{\nu}_p}{4}\left(1 - \frac{k_B}{\mu_p^2 m_p c^2}\widehat{T}_p\right)$$

$$\widehat{M}_{jk\ell}^p = A_p(B_p)^3 = \frac{\widehat{\nu}_p}{16}\left(\frac{3k_B}{\mu_p^2 m_p c^2}\widehat{T}_p - 1\right)$$

$$(2.38)$$

independently on the velocity indexes i, j, k, ℓ.

2.2.4   Shizuta–Type Models

Recently, in papers [9–11] Shizuta and coworkers proposed a new methodology to construct spatial models with a large number of velocities, which consider binary collision only and have the right dimension of the space of collisional invariants.

For instance, in [11] Shizuta and Kawashima proved that for any integer k there exists a regular discrete velocity model with k moduli. The velocity discretization presented is invariant under rotation of $\frac{\pi}{2}$ around the coordinate axis and under central inversion.

The model can be constructed as follows. Consider the cube designed in Fig. 2.6 and define the following sets of constant vectors with the same modulus, which for simplicity will be called sub-discretizations

$$\mathcal{V}_1 = \{\mathbf{v}_i^{(1)}, i = 1, \ldots, 6 \text{ joining } \mathbf{C} \text{ with the centers of the faces}\} \;,$$

$$\mathcal{V}_2 = \{\mathbf{v}_i^{(2)}, i = 1, \ldots, 12 \text{ joining } \mathbf{C} \text{ with the centers of the edges}\} \;,$$

$$\mathcal{V}_3 = \{\mathbf{v}_i^{(3)}, i = 1, \ldots, 6 \text{ joining } \mathbf{C} \text{ with the vertices}\} \;.$$

It can be proved by inspection of the collision mechanics that the velocity discretizations

$$\mathcal{V}_1 \;, \quad \mathcal{V}_1 \cup \mathcal{V}_2 \;, \quad \mathcal{V}_1 \cup \mathcal{V}_2 \cup \mathcal{V}_3$$

correspond to regular models, in particular $\dim \mathcal{M}$ is always equal to five. Note that $\mathcal{V}_1 \cup \mathcal{V}_3$ is the Cabannes' 14–velocity model.

If more velocity moduli are desired, then one has to consider the sub-discretizations

$$\mathcal{V}_4 = \{\mathbf{v}_i^{(4)}, i = 1, \ldots, 6 \; : \; \mathbf{v}_i^{(4)} = 2\mathbf{v}_i^{(1)}\} \;,$$

$$\mathcal{V}_5 = \{\mathbf{v}_i^{(5)}, i = 1, \ldots, 24 \text{ which can be obtained by rotations of}$$

$$\text{the vector } (2c, c, c) \text{ of multiples of } \frac{\pi}{2} \text{ around the axis}\} \;,$$

$$\mathcal{V}_6 = \{\mathbf{v}_i^{(6)}, i = 1, \ldots, 12 \; : \; \mathbf{v}_i^{(6)} = 2\mathbf{v}_i^{(2)}\} \;,$$

and recursively

$$\mathcal{V}_k = \{\mathbf{v}_i^{(k)} \; : \; \mathbf{v}_i^{(k)} = 2\mathbf{v}_i^{(k-4)}\} \, .$$

It can be shown that any velocity discretization $\bigcup_{i=1}^{k} \mathcal{V}_i$ defines a model with $\dim \mathcal{M} = 5$, so that the Maxwellians can be written as in (1.49).

Following the methodology given in [11], one can construct regular planar models with a large number of velocities and with the property that $\dim \mathcal{M} = 4$.

These can be obtained by considering the following sub-discretizations

$$\mathcal{V}_1 = \{c(1,0)\, , \, c(0,1)\, , \, c(-1,0)\, , \, c(0,-1)\} \, ,$$

$$\mathcal{V}_2 = \{c(1,2)\, , \, c(2,1)\, , \, c(-2,1)\, , \, c(1,-2)\, ,$$

$$c(-1,-2)\, , \, c(-2,-1)\, , \, c(2,-1)\, , \, c(1,-2)\} \, ,$$

$$\mathcal{V}_3 = \{\mathbf{v}_i^{(3)} \; : \; \mathbf{v}_i^{(3)} = 3\mathbf{v}_i^{(1)}\} \, ,$$

and recursively

$$\mathcal{V}_{2k+1} = \{\mathbf{v}_i^{(2k+1)} \; : \; \mathbf{v}_i^{(2k+1)} = (2k+1)\mathbf{v}_i^{(1)}\} \, ,$$

$$\mathcal{V}_{2k+2} = \{(2k+1,2)\, , \, c(2,2k+1)\, , \, c(-2,2k+1)\, , \, c(2k+1,-2)\, ,$$

$$c(-2k-1,-2)\, , \, c(-2,-2k-1)\, , \, c(2,-2k-1)\, , \, c(2k+1,-2)\}.$$

In a way similar to the one used in [11], it can be proved that every model with velocity discretization equal to $\bigcup_{i=1}^{k} \mathcal{V}_i$ has $\dim \mathcal{M} = 4$.

References

[1] R. Gatignol, **Théorie Cinétique des Gaz a Répartition Discrète de Vitesses**, *Lecture Notes in Phys. No.***36**, Springer-Verlag, Berlin, New York, 1975.

[2] R. Gatignol, "Unsteady Couette flow for a discrete velocity gas", in **Rarefied Gas Dynamics**, Ed. R. Campargue, CEA, Paris, 1978, p.195.

[3] E. Longo, "Analytical solutions of the discrete Boltzmann equation for the steady Couette flow for binary gas mixtures", *Mech. Res. Comm.*, **10**, 1983, p.373.

[4] E. Longo and R. Monaco, "On the discrete kinetic theory with multiple collisions: A plane six velocity model and unsteady Couette flow", in **Rarefied Gas Dynamics: Theoretical and Computational Techniques**, Eds. E.P. Muntz, D.H. Campbell and D.P. Weaver, AIAA Publ. vol. **118**, Washington, 1989, p.118.

[5] I.E. Broadwell, "Shock structure in a simple discrete velocity gas", *Phys. Fluids*, **7**, 1964, p.1243.

[6] I.E. Broadwell, "Study of rarefied shear flow by the discrete velocity method", *J. Fluid Mech.*, **19**, 1964, p.401.

[7] H. Cabannes, "Etude de la propagation des ondes de choque dans un gaz a 14 vitesses", *J. de Mecanique*, **14**, 1975, p.705.

[8] E. Longo and R. Monaco, "On the thermodynamics of the discrete models of the Boltzmann equation for gas mixtures", *Transp. Theory Statist. Phys.*, **17**, 1988, p.423.

[9] Y. Shizuta and S. Kawashima, "System of equations of hyperbolic-parabolic type with application to the discrete Boltzmann equation", *Hokkaido Math. J.*, **14**, 1985, p.249.

[10] Y. Shizuta and S. Kawashima, "The regularity of discrete models of the Boltzmann equation", *Proc. J. Acad., Ser. A*, **61**, 1985, p.252.

[11] Y. Shizuta and S. Kawashima, "The regular discrete models of the Boltzmann equation", *J. Math. Kyoto Univ.*, **27**, 1987, p.131.

CHAPTER 3

ON THE DISCRETE BOLTZMANN

EQUATION IN UNBOUNDED DOMAINS

This chapter deals with the solution and simulation of the discrete Boltzmann equation in unbounded domains. This topic can be regarded as a preliminary step before approaching specific problems of fluid-dynamics, which will be dealt with in the following chapters.

Referring to the solution of the initial value problem, it is important to have in the background suitable existence (and uniqueness) theorems which may hopefully support the application of numerical algorithms to the solution of specific problems.

Even if one has to be aware that an existence theorem for the solutions to the initial value problem is still far from an existence theory for specific problems in fluid-dynamics, nevertheless it is useful to have a detailed knowledge on the matter both towards the validity of the models and the application of numerical techniques.

Considering that this volume is mainly concerned with applications, details of mathematical proofs will not be given. However we shall provide a rather complete survey and details of the functional spaces where the mathematical proofs are obtained. This according to the fact that numerical algorithms and analytic simulations need to be related to suitable functional spaces, then the solution has to be proved to exist in the said spaces.

The chapter is organized in two parts. The first section gives the mathematical formulation of the initial value problem in unbounded domains and a survey of existence results. The second section deals with the quantitative analysis, that is with the solution of the discrete Boltzmann equation by application of suitable numerical algorithms or by simulation schemes. While the first section is organized at a survey level, the second section can be regarded as an introduction to the quantitative analysis of the discrete Boltzmann equation in a line which will be developed further in the chapters which follows.

3.1   The Cauchy Problem for the Discrete Boltzmann Equation

We consider in this section some mathematical aspects of the initial value problem in unbounded domains (the Cauchy problem) for the discrete Boltzmann equation. The section is organized in three sub-sections. First we deal with the mathematical formulation of the problem, with the definitions of solutions and with an existence, local in time, theorem. Then we provide a survey of the mathematical problem with initial data in the whole space $\mathbb{R}^d$, $d \geq 2$. Finally a survey on the mathematical analysis for problems in one space dimensions is given. The aim of the review is to provide a useful background for numerical applications and specific simulations.

3.1.1 Mathematical Formulation and Local Existence

Consider a discrete velocity model of the Boltzmann equation

$$\frac{\partial N_i}{\partial t} + \mathbf{v}_i \cdot \nabla_{\mathbf{x}} N_i = J_i[\mathbf{N}] \tag{3.1a}$$

where J_i can include both binary and triple collisions.

The initial value problem in the whole space $\mathbb{R}^d$, $d \geq 2$ is defined by Eq.(3.1a) joined to the initial data

$$N_i(t = 0, \mathbf{x}) = N_{io}(\mathbf{x}) \ , \quad i = 1, \ldots, n \ , \quad \forall \mathbf{x} \in \mathbb{R}^d \tag{3.1b}$$

that is

$$\mathbf{N}(t = 0, \mathbf{x}) = \mathbf{N}_o(\mathbf{x}) \ , \quad \mathbf{N} = \{N_1, \ldots, N_n\} \quad . \tag{3.1c}$$

The solution of the Cauchy problem is a function set $N_i(t, \mathbf{x})$, with each N_i positive, satisfying, in a suitable functional space $\mathcal{B}$, Eq.(3.1a) and the initial conditions (3.1b). Details on the function spaces $\mathcal{B}$ which can be used for these problems will be given in what follows.

The initial value problem can be written in integral form. Classically this form is obtained integrating the differential equation (3.1a) along the characteristic lines

$$\mathbf{x}^{\#} = \mathbf{x} + \mathbf{v}_i t \quad .$$

(3.2)

Consequently one can define

$$N_i^{\#}(t, \mathbf{x}) = N_i(t, \mathbf{x}^{\#})$$
$$J_i^{\#}[\mathbf{N}(t, \mathbf{x})] = J_i[\mathbf{N}(t, \mathbf{x}^{\#})]$$

(3.3)

and the initial value problem takes the form

$$i = 1, \ldots, n \ : \ \ N_i^{\#}(t, \mathbf{x}) = N_{io}(\mathbf{x}) + \int_0^t J_i^{\#}[\mathbf{N}(s, \mathbf{x})] \, ds \ .$$

(3.4)

Another integral form which is commonly used in discrete kinetic theory can be obtained observing that the differential equation (3.1a) can be re-written, using (3.2–3.3), as follows

$$\frac{dN_i^{\#}}{dt} + N_i^{\#} R_i^{\#}(\mathbf{N}) = G_i^{\#}(\mathbf{N}) \ ,$$

(3.5)

where $G_i^{\#}$ and $N_i^{\#} R_i^{\#}$ are the gain and loss terms, respectively, and where

$$R_i^{\#}(\mathbf{N}) = \frac{1}{2} \sum_{jhk} A_{ij}^{hk} N_j^{\#} \ .$$

Then the following exponential integral form can be derived

$$
N_i^{\#}(t, \mathbf{x}) = N_{io}(\mathbf{x}) \exp\left[-\int_0^t R_i^{\#}(\mathbf{N})(s, \mathbf{x}) \, ds \right]
$$
$$
+ \int_0^t G_i^{\#}(\mathbf{N})(s, \mathbf{x}) \exp\left[-\int_s^t R_i^{\#}(\mathbf{N})(r, \mathbf{x}) \, dr \right] ds \tag{3.6}
$$

for $(i = 1, \ldots, n)$.

Some definitions can now be given

Definition 3.1: *A function set* $\mathbf{N}(t, \mathbf{x})$ *which is positive in* $\mathcal{B}$ *and satisfies Eq.(3.4) or Eq.(3.6) is called a mild solution.*

Definition 3.2: *A function set* $\mathbf{N}(t, \mathbf{x})$ *which is positive in* $\mathcal{B}$ *and continuously differentiable with respect to* $\mathbf{x}$ *and* t *fulfilling Eq.(3.4) or Eq.(3.6) is called classical solution.*

Definition 3.3: *A function set* $\mathbf{N}(t, \mathbf{x})$ *which satisfies Eq.(3.5) with initial data* $\mathbf{N}_o^{\#} = \mathbf{N}_o$, *is positive in* $\mathcal{B}$ *and strongly differentiable, is called strong solution.*

Of course $\mathcal{B}$ will not always be the same functional space and must be specified in each particular case of existence theorems. Examples of function spaces useful for the problems are given in the definitions which follow

Definition 3.4: *Let* B^m *be the set of all functions with* m*-derivatives continuous and bounded in* $\mathbb{R}^d$. *Moreover* $\mathbf{B}^m = (B^m)^n$; *then if*

$\gamma = \{\gamma_1, \ldots, \gamma_b\}$ *is a multi-index and* $|\gamma| = \gamma_1 + \cdots + \gamma_b$, *the norm in* $\mathbf{B}^m$ *is defined by*

$$\|\mathbf{N}\|_m = \max_{i \le n} \sup_{\substack{|\gamma| \le m \\ \mathbf{x} \in \mathbf{R}^d}} \left| \frac{\partial^{|\gamma|} N_i}{\partial \mathbf{x}^\gamma} \right| < \infty \ .$$

Definition 3.5: *Let* $C^s(0, T; B^m)$ *be the set of all s–times continuously differentiable functions from* $[0, T]$ *into the Banach space* B^m. *Then* $C^s(0, T; \mathbf{B}^m) = (C^s(0, T; B^m))^n$ *and the norm in* $C^s(0, T; \mathbf{B}^m)$ *is given by*

$$\|\|\mathbf{N}\|\|_m = \sup_{\substack{t \in [0,T] \\ k \le s}} \left\| \frac{d^k \mathbf{N}}{dt^k} \right\|_m < \infty \ .$$

Definition 3.6: *Let* $L^\infty(0, T; B^m)$ *be the set of all essentially bounded functions from* $[0, T]$ *into* B^m; *moreover the norm in*

$$L^\infty(0, T; \mathbf{B}^m) = (L^\infty(0, T; B^m))^n$$

will be denoted by

$$\|\|\mathbf{N}\|\|_m^\infty = \operatorname*{ess\,sup}_{t \in [0,T]} \|\mathbf{N}(t)\|_m < \infty \ .$$

Furthermore the subscript $_+$ will denote the set of positive functions.

After these preliminaries, all elements to state a local existence theorem have been given. Obtaining a local existence theorem, namely an existence theorem for a limited time interval, is a matter of application of classical methods of functional analysis. As a matter of fact local existence has been proven by several authors. Even though these results are not reviewed here, one has to keep in mind that existence theorems are certainly useful for the applications if the maximal estimate of the existence result and all information on the regularity of the solutions are provided. This allows to safely apply solution algorithms, as we shall see in what follows.

Since this book is addressed to the analysis of fluid-dynamic problems and not to existence theory, the reader who is interested in a deeper insight into such a problem can find more complete indications in the review papers by Platkowski and Illner [1] and by Bellomo and Gustafsson [2].

With this in mind, the following (among others) result can be given

Theorem 3.1   ([3])

Let $\mathbf{N}_o \in \mathbf{B}_+^m$ *with* $m \geq 2$ *and* $\mathbf{x} \in \mathbb{R}^d$ *, and let*

$$\mathcal{T}_c = \left[8 \max_{i \leq n} \sum_{jhk} A_{ij}^{hk} \|\mathbf{N}_o\|_m \right]^{-1} .$$

Then there exists a unique positive mild solution to Eq.(3.4) in the space

$\mathbf{B}^m$ on the *time interval* $[0, \mathcal{T}_c]$ *with*

$$\mathbf{N} \quad \in L^\infty(0, \mathcal{T}_c; \mathbf{B}^m)_+ \cap \mathcal{C}^\circ(0, \mathcal{T}_c; \mathbf{B}^{m-1})_+ \cap \mathcal{C}^1(0, \mathcal{T}_c; \mathbf{B}^{m-2})$$

$$\mathbf{N}^\# \in \mathcal{C}^\circ(0, \mathcal{T}_c; \mathbf{B}^m)_+ \cap \mathcal{C}^1(0, \mathcal{T}_c; \mathbf{B}^{m-1})$$

$$\frac{d\mathbf{N}^\#}{dt} \in L^\infty(0, \mathcal{T}_c; \mathbf{B}^m) \ . \qquad\qquad\qquad \blacksquare$$

The proof is obtained by application of the classical fixed point theorem [4] to the integral form (3.4) which may be written, in operator form, as

$$\mathbf{N} = \mathcal{U}\mathbf{N} \ . \tag{3.7}$$

It is proven in [3] that if $t \le \mathcal{T}_c$ the operator $\mathcal{U}$ is contractive in a closed subset of $\mathbf{B}^m_+$. Positivity is proved exploiting the integral form (3.6).

The proof of regularity is obtained, as usual, by analysing the evolution equation for the space derivative of $\mathbf{N}$ and applying the same fixed point technique to such an equation.

A result, analogous to the one of Theorem 3.1 was proven for discrete velocity model with triple collisions in [5].

Starting from the existence Theorem 3.1, which holds locally in time, namely for $t < \mathcal{T}_c$, a brief review of the theorems on global existence of solutions will be given.

The survey is organized for different classes of problems and will essentially consists in a qualitative description of the result and in the related bibliographical indication.

3.1.2 *Existence of Solutions in the Whole Space* $\mathbb{R}^d$, $d \geq 2$

Existence of solutions in the whole space is known, in the literature, only under suitable smallness assumptions on the size, in norm, of the initial data. Therefore the mathematical results for the Cauchy problem in $\mathbb{R}^d$ for the discrete kinetic theory are analogous to the ones described in the book [6] for the full Boltzmann equation and are essentially referred to the case of small perturbations of vacuum or of Maxwellian equilibrium.

Existence results for *large* initial data such as the ones obtained for the full Boltzmann equation by DiPerna and Lions [7] seem, at present, not applicable to the discrete Boltzmann equation.

Limiting our attention to general discrete models (stronger results can be obtained for special cases), one needs mentioning the existence result by Toscani [8] who obtained global existence for *small*, in norm, initial conditions decaying exponentially to zero at infinity. This result is generalized in [9] to the case of inverse polynomial decaying to zero at infinity in space.

Bony [10] proved global existence for small initial data without decay hypothesis.

In all cases the proof is based upon contractivity properties of the operator $\mathcal{U}$ defined in (3.7). The results of [8–10] refer to the case of the discrete Boltzmann equation with binary collisions only. Paper [5] generalizes the result of [9] to the solution of the model with triple collision term.

The results reported until now refer to small, in norm, perturbation of vacuum. There exist existence results for the solution to the initial value problem in the whole space for perturbations of a Maxwellian equilibrium state [11,12].

The line to obtain these results consists in the following three steps

i) Definition of the Maxwellian equilibrium state and of the space of collisional invariants $\mathcal{M}$;

ii) Statement of suitable stability conditions based upon properties of $\mathcal{M}$;

iii) Proof of global existence, uniqueness and trend to equilibrium for small perturbations of the Maxwellian state.

Once more these results are conditioned by some smallness assumption on the initial data: essentially on the distance of the perturbation from equilibrium.

3.1.3   *The Initial Value Problem in One-Space Dimension*

The Cauchy problem in one dimension provides somewhat *stronger* results than the one in the whole space. In fact in several cases one can remove *smallness* assumptions on the initial data.

The first result on this topic was given by Nishida and Mimura [13] for the 6–velocity Broadwell model in one dimension (see Section 2.2.1) which, introducing the dimensionless variables

$$\eta = \frac{x}{\ell} \qquad \tau = \frac{ct}{\ell} \qquad N_i' = \frac{N_i}{N} , \qquad\qquad (3.8)$$

where ℓ is a fixed laboratory length scale and $\mathcal{N} = \frac{3}{4S\ell}$, can be written as

$$\frac{\partial N'_1}{\partial \tau} + \frac{\partial N'_1}{\partial \eta} = J(\mathbf{N}')$$

$$\frac{\partial N'_2}{\partial \tau} = -\frac{1}{2}J(\mathbf{N}') \qquad (3.9)$$

$$\frac{\partial N'_3}{\partial \tau} - \frac{\partial N'_3}{\partial \eta} = J(\mathbf{N}')$$

with $J(\mathbf{N}') = (N'_2)^2 - N'_1 N'_3$.

The existence proof by Nishida and Mimura consists first in giving a local existence theorem and then, using an a priori estimate, in extending the result globally in time for small initial data. This result will be reported here because of its historical relevance. It is, in fact, the first global existence result for the discrete Boltzmann equation.

Theorem 3.2   ([13])
Let $\mathbf{N}'_o \in \left(B^2(\mathbf{R}) \cap L^1(\mathbf{R})\right)^3_+$ with $\max_{i \leq 3} N'_{io} < M_o$, and

$$\nu_o = \int_{-\infty}^{+\infty} [N'_{10}(\eta) + 4N'_{20}(\eta) + N'_{30}(\eta)]\, d\eta < 4 .$$

Then there exists a unique global classical solution to the Cauchy problem for Eq.(3.9) with initial data $\mathbf{N}'_o$ and

$$\forall \tau \in [0, \infty) , \quad \eta \in \mathbf{R} : \quad N'_i(\tau, \eta) \leq \frac{8M_o}{4 - \nu_o} , \quad i = 1, 2, 3 . \qquad \blacksquare$$

Technical generalizations of this theorem were investigated by Cabannes [14,15] for his 14–velocity model and for the planar regular $2n$–velocity model.

An important improvement to Theorem 3.2 was given by Tartar [16,17], again for the one-dimensional Broadwell model (3.9), who proved global existence for large initial data.

Tartar's result is of relevant importance. In fact he was able to exploit the decay in time of the H–functional to obtain global existence for *large* and space-periodic initial data with L_1 and L^∞ properties.

This result has been the starting point of several further studies of other authors: Beale [18,19], Bony [20], Cabannes and Kawashima [21] and Toscani [22]. All these papers are organized in order to generalize Tartar's result and to obtain suitable convergence properties. Referring to the review papers [1,2] for details on the technical aspects of the papers which have been cited above, it is worth mentioning that the relevant aspect of the results of [18–22] is that Tartar's theorem has been extended to rather general discrete velocity models of the Boltzmann equation with convergence properties. The crucial point of the problem may be stated as the possibility of obtaining similar existence proofs in three space dimensions, hopefully exploiting other Liapunov functionals. Such a problem is still open and is a challenging topic for applied mathematicians.

3.2 On the Solution of the Discrete Boltzmann Equation

The solution to the initial value problem for the discrete Boltzmann equation can be hopefully obtained in analytic form. This is the case of some analytic solutions obtained by various authors for classical fluid-dynamic problems (shock waves, Couette and Rayleigh flows). These results, as reviewed in [1], have been obtained for simple models, for instance the planar or spatial Broadwell model, both for a simple gas and for binary mixtures.

On the other hand, the aim of the present book is to deal with very general physical situations, say models with a large number of velocities, models for multicomponent gas mixtures, initial-boundary value problems with very general boundary conditions and so on. Since it does not seem, at least at present, that analytic solution can be obtained in this general framework, we will only mention here an interesting approach developed by Cornille in various papers (see [23,24] and the bibliography therein quoted) and by other authors (see among others Cabannes and Thiem [25] and Cabannes [26]). Such a method consists in postulating the form of the solution and in determining the parameters characterizing the solution using the conservation equations.

In particular the solution can be sought for in soliton-type form

$$N_i(z_i) = a_i f_i(z_i) \quad , \quad z_i = \Lambda_i x + \Omega_i t \quad , \quad i = 1, \dots, n \qquad (3.10)$$

$$a_i \in \mathbb{R}_+ \quad , \quad \Lambda_i, \Omega_i \in \mathbb{R}$$

or in bisoliton-type form

$$N_i(t, x) = a_i + 2 \, \mathcal{R}e\{b_i \tan(\Lambda x + \sqrt{-1} \, \Omega t)\} \quad , \quad i = 1, \dots, n \qquad (3.11)$$

$$a_i, \Lambda, \Omega \in \mathbb{R} \quad , \quad b_i \in \mathbb{C}$$

where $\mathbb{C}$ is the set of complex numbers. The problem is then to determine the constants which appear in (3.10,3.11).

Also in this case, it does not seem immediate to obtain analytic solution for very general physical situations and for specific problems in fluid-dynamics. Therefore the solution of problems has to be, finally sought by numerical schemes.

As we have mentioned several times, the quantitative analysis of the discrete Boltzmann equation can be obtained by quite standard numerical methods. This is one of the great advantages with respect to the continuous models of nonlinear kinetic theory. As usual, one needs the support of an existence theorem consistent with the solution scheme.

The quantitative solution can be obtained, for instance, by application of the *differential quadrature method* [27] or of *finite difference methods*. The applications dealt with in this volume will use these two methods.

The first method will be described in its main lines in this section, the reader is referred to the classical literature of numerical analysis, see for instance [28,29], for the application of finite differences schemes.

The first step in the application in one space dimension of the differential quadrature method consists in interpolating the solution (in our case of the initial value problem) as follows

$$N_i(t, x) \approx \sum_{r=1}^{R} L_r(x) N_{ir}(t) \quad , \tag{3.12}$$

where

$$N_{ir}(t) = N_i(t; x = x_r)$$

and the $L_r(x)$ are the Lagrange interpolation polynomials

$$L_r(x) = \prod_{\substack{k=1 \\ k \neq r}}^{R} \frac{x - x_k}{x_r - x_k} \ . \tag{3.13}$$

Moreover the spatial derivatives of N_i are approximated by

$$\frac{\partial N_i}{\partial x}(t; x = x_r) \approx \sum_{s=1}^{R} a_{rs} N_{is}(t) \tag{3.14}$$

where

$$a_{rs} = \left. \frac{dL_s}{dx} \right|_{x=x_r} \ . \tag{3.15}$$

Standard calculations yield

$$a_{rs} = \frac{L'_*(x_r)}{(x_r - x_s)L'_*(x_s)} \ , \quad r \neq s \tag{3.16}$$

and

$$a_{rr} = \frac{L''_*(x_r)}{2L'_*(x_r)} \ , \tag{3.17}$$

where

$$L_*(x) = \prod_{k=1}^{R} (x - x_k) \ .$$

Substituting the various expressions (3.12–3.17) yields, as we shall see in details for a specific model, a system of ordinary differential equations of the type

$$\frac{dN_{ir}}{dt} = J_i[\mathbf{N}_r] - \sum_{s=1}^{R} a_{rs} N_{is}$$

$$N_{iro} = N_{ir}(t = 0) ,$$

(3.18)

where $\mathbf{N}_r = \{N_{1r}, \ldots, N_{nr}\} \in \mathbf{R}^n$.

This method will be used, in the chapters which follow, to solve fluid-dynamic problems. As a matter of fact, this method has the advantage that the boundary conditions can be imposed in a straightforward fashion. In fact, one has simply to replace a suitable set of those equations of (3.18) which correspond to the boundaries (say for $x = x_1$ and $x = x_R$) with the boundary conditions themselves. Therefore, even if we shall use this method here for a purely initial value problem, this method is suggested, and will be applied in Chapter 5, for initial-boundary value problems.

Within the framework of this chapter, then essentially at the level of numerical experiments, the method will be tested in the framework of the existence results surveyed in Section 3.1. The test refers to a six velocity model with initial conditions decaying to zero at infinity.

The calculations have to be regarded as a simple experiment, the reader may perform analogous calculations in two space dimensions or referred to more general models. This suggestion can be regarded as a proposed exercise to become more acquainted with numerical calculations which will be developed further in the chapters which follow.

Considering that the experiment is in one space dimension, one should expect that the initial value problem always has a solution for L_1 data. In fact conservation of mass and momentum and the application of the H–theorem assure global existence without smallness assumptions [16–22].

On the other hand, if one deals with the problem for a discrete velocity model such that the loss term is eliminated, then global existence should be expected only for small initial data. In fact, in this case, the problem is analogous to the solution in the whole space, which is such that the H–theorem cannot be used.

The numerical experiments will confirm, as we shall see, the qualitative behaviour which has been described above.

Keeping this in mind and in order to show, at a practical level, the applicability of the differential quadrature method, consider the 6– velocity planar model with binary collisions only, described in Section 2.1.3. In one space dimension

$$\frac{\partial \mathbf{N}}{\partial y} = 0, \qquad N_2 = N_6 \quad \text{and} \quad N_3 = N_5$$

and then Eq.(2.12) can be written as

$$\begin{aligned}
\frac{\partial N_1}{\partial t} + c\frac{\partial N_1}{\partial x} &= J(\mathbf{N}) \\
\frac{\partial N_2}{\partial t} + \frac{c}{2}\frac{\partial N_2}{\partial x} &= -\frac{1}{2}J(\mathbf{N}) \\
\frac{\partial N_3}{\partial t} - \frac{c}{2}\frac{\partial N_3}{\partial x} &= -\frac{1}{2}J(\mathbf{N}) \\
\frac{\partial N_4}{\partial t} - c\frac{\partial N_4}{\partial x} &= J(\mathbf{N})
\end{aligned} \tag{3.19}$$

with $J(\mathbf{N}) = \frac{4}{3}cS(N_2N_3 - N_1N_4)$.

This system can be written in dimensionless form by introducing the variables

$$\tau = \frac{ct}{\ell} \quad , \quad \zeta = \frac{x}{\ell} \quad , \quad N_i' = \frac{N_i}{\mathcal{N}} \quad ,$$

where ℓ is again a fixed laboratory length scale and $\mathcal{N} = \frac{3}{4S\ell}$.

Consider the initial value problem with

$$N_{io}'(z) = \frac{\beta_i}{1 + \zeta^2} \quad , \quad i = 1, \ldots, 4 \ ,$$

where β_i are constants.

To integrate numerically Eqs.(3.19) one can use the collocation scheme which has been described above. The space variable $\zeta \in \mathbf{R}$ is rescaled in a bounded interval by the transformation

$$\eta = \frac{e^\zeta}{1 + e^\zeta} \quad , \quad \zeta \in \mathbf{R} \quad , \quad \eta \in (0,1) \ .$$

Then the initial data become

$$N_{io}'(\eta) = \frac{\beta_i}{1 + \log^2 \frac{\eta}{1-\eta}} \quad , \quad i = 1, \ldots, 4 \ . \tag{3.20}$$

Moreover if the variable η is discretized into R nodal points η_r , $r = 1, \ldots, R$ in the interval $(0,1)$, then the system of differential equations can be re-written in the following integral form

$$\mathbf{N}_{ir}'(\tau) = \frac{\beta_i}{1 + \log^2 \frac{\eta_r}{1-\eta_r}} - \eta_r(1-\eta_r)v_i \int_0^\tau \sum_{s=1}^R a_{rs} N_{is}'(\xi)\, d\xi$$

$$+ \int_0^\tau J_i(\mathbf{N}')(\xi; \eta = \eta_r)\, d\xi \ , \tag{3.21}$$

where

$$\{v_i\} = \{1, \frac{1}{2}, -\frac{1}{2}, -1\} \quad , \quad \{J_i\} = \{J, -\frac{J}{2}, -\frac{J}{2}, J\} .$$

The results are shown in Figs. 3.1 and 3.2 where the rescaled numerical density

$$\nu'(\tau, \eta) = N_1'(\tau, \eta) + 2N_2'(\tau, \eta) + 2N_3'(\tau, \eta) + N_4'(\tau, \eta)$$

is plotted versus space at different times. In Fig.3.1, Eqs.(3.21) are integrated with $\beta_2 = \beta_3 = \beta_4 = 0.001$, while $\beta_1 = 3$ (Fig. 3.1a) and $\beta_1 = 4$ (Fig. 3.1b).

As can be seen, the result is a travelling wave of soliton-type which propagates to the right. Note that the maximum value of this soliton decreases in both cases.

In Fig. 3.2 a similar experiment is performed with the difference that in the collisional term only the gain is computed (the loss term is put equal to zero). It can be observed that for $\beta_1 = 3$ (Fig. 3.2a) the behaviour is analogous to that of Fig. 3.1a, whereas for $\beta_1 = 4$ (Fig. 3.2b) the numerical density suddenly blows-up.

The quantitative behaviour shown by the numerical calculations is in agreement with the existence theory surveyed in Section 3.1. The reader can develop further these calculations in order both to practice with numerical analysis which will be useful for the quantitative analysis of fluid-dynamic problems and to test the existence theorems surveyed in the preceding section.

The following exercises, among others, are suggested

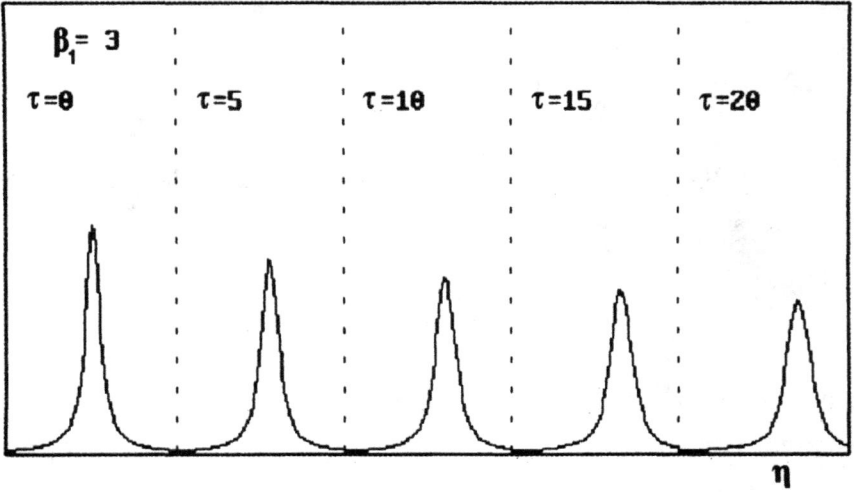

(a)

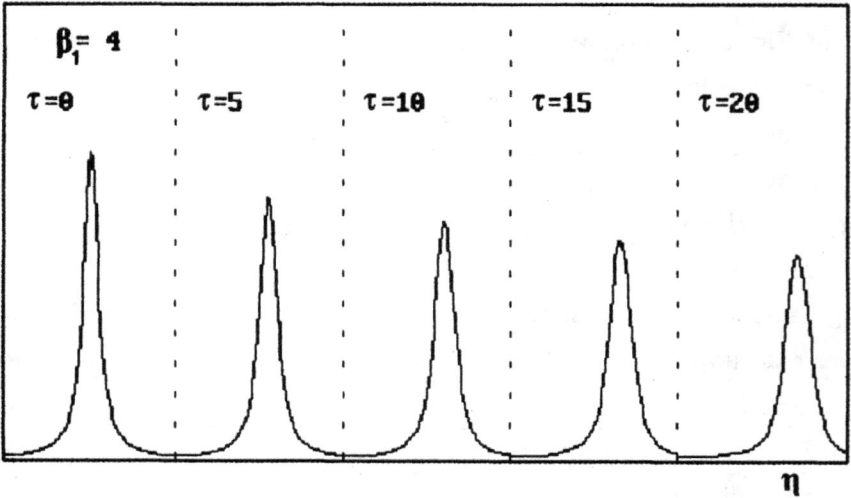

(b)

Fig. 3.1 — *Evolution of the density vs space for the full equation*

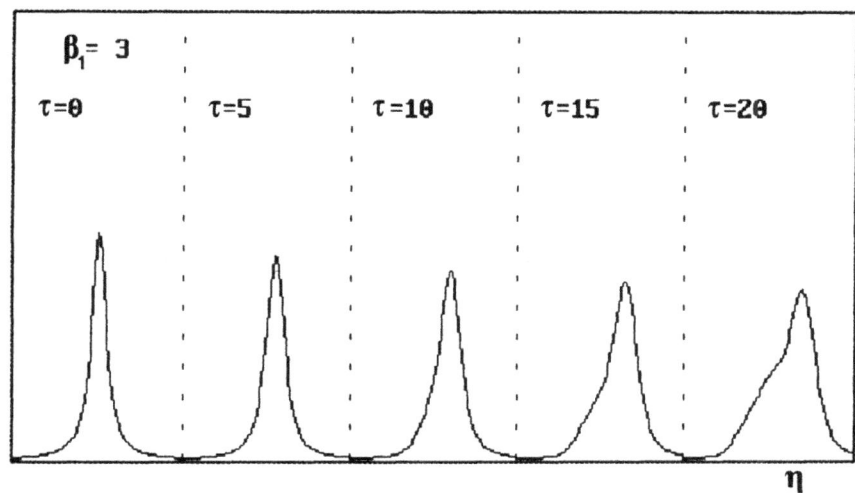

(a)

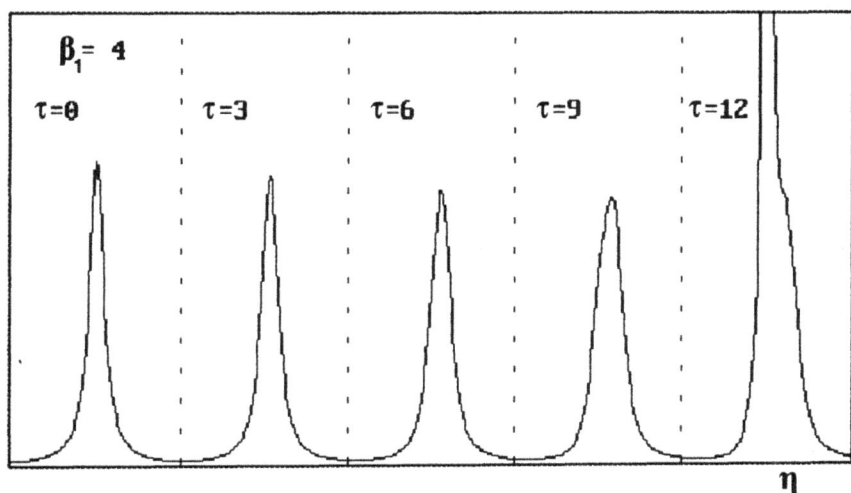

(b)

Fig. 3.2 — *Evolution of the density vs space for the equation with the gain term only*

– Analysis of the same model in two space dimensions,

– Analysis of the same problem, in one or more than one space dimensions, for discrete velocity models more general that the one which have been dealt with above. In particular, one can show how increasing the number of velocities may affect quantitatively the number densities profiles.

Referring to problems in two space dimensions, the calculations can be performed in a fashion analogous to the one indicated in Eqs.(3.12–3.17). The starting point is the interpolation of the densities first in one direction

$$N_i(t, x, y) \approx \sum_{r=1}^{R} L_r(x) N_{ir}(t, y) \tag{3.22a}$$

and then in the other

$$N_{ir}(t, y) \approx \sum_{h=1}^{H} L_h(y) N_{irh}(t) , \tag{3.22b}$$

so that $N_i(t, x, y)$ is interpolated as

$$N_i(t, x, y) \approx \sum_{r=1}^{R} \sum_{h=1}^{H} L_r(x) L_h(y) N_{irh}(t) , \tag{3.23}$$

where L_r are defined by Eq.(3.13) and

$$N_{irh}(t) = N_i(t, x_r, y_h) .$$

Following the same procedure used in the one-dimensional case allows to obtain the expressions of the space derivatives in the nodal points and finally a system of RH ordinary differential equations.

The content of this chapter can be completed reporting, briefly, an alternative approach recently developed for discrete models of fluid-dynamics.

The method developed by Gabetta [30] consists first in discretizing the space variables into a suitable network of nodal points (in a fashion similar to the one required for lattice fluid-dynamics), which defines the velocity discretization, then the flow exchanges in the nodal points are defined by suitable transition probabilities consistent with mass and momentum conservation.

Even if this type of modelling, at present, is still referred to spatially homogeneous cases for very simple velocity schemes and still needs to be developed in its mathematical theory, it is worth reporting some calculations in order to show the practical application to flow pattern simulations.

Once the theory and the modelling will be suitably developed, the simulation scheme may, hopefully, become an alternative to the direct integration of the discrete Boltzmann equation.

With this in mind, consider the 4–velocity Gatignol model (see Eqs.(2.4)) with velocities $\mathbf{v}_k$ parallel to the axes of the plane fixed orthogonal frame $O(x,y)$

$$\frac{\partial N_1}{\partial t} + c\frac{\partial N_1}{\partial x} = J(\mathbf{N})$$

$$\frac{\partial N_2}{\partial t} + c\frac{\partial N_2}{\partial y} = -J(\mathbf{N})$$

$$\frac{\partial N_3}{\partial t} - c\frac{\partial N_3}{\partial x} = J(\mathbf{N}) \qquad (3.24)$$

$$\frac{\partial N_4}{\partial t} - c\frac{\partial N_4}{\partial y} = -J(\mathbf{N})$$

with $J(\mathbf{N}) = cS(N_2 N_4 - N_1 N_3)$.

Let us now discretize the plane xy localizing the knots (i,j) in an indefinite planar lattice with cells of width $c\Delta t$, so that any change of directions of particles, due to collisions, can occur only in the knots every instant Δt. Then at each instant $n\Delta t$ and at each knot (i,j), a random variable $\alpha_{i,j}^n$ can be defined such that it can assume only five discrete different numerical values $k = 0,1,2,3,4$. In particular $k = 0$ means that no particle is present at that time in the knot (i,j). The other four values are the ones joined to presence of particles with velocity $\mathbf{v}_k$, $k = 1,2,3,4$. Finally a probability measure

$$P(\alpha_{i,j}^n = k) \quad , \quad n,i,j,k \text{ integers}$$

can be defined, which gives the probability that, after a time $n\Delta t$ from the initial time $n = 0$ in the knot (i,j) there is a particle with velocity $\mathbf{v}_k$ (or no particles). Each event in the (i,j)–knot will depend then on the state of the four adjacent knots (see the picture below)

○   ○   ●   ○   ○

$$(i, j+1)$$

○   ●   ●   ●   ○

$$(i-1, j) \qquad (i, j) \qquad (i+1, j)$$

○   ○   ●   ○   ○

$$(i, j-1)$$

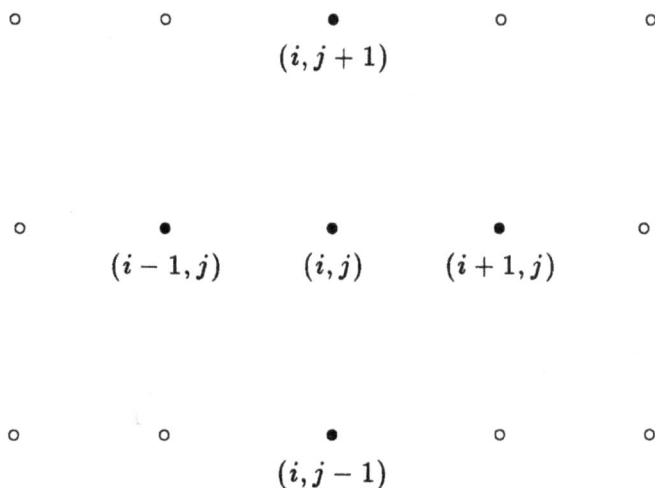

Fig. 3.3 — *Mesh of the lattice gas theory*

at time $(n-1)\Delta t$ so that at time $n\Delta t$ in (i, j) we will have particles
with velocities provided by the products due to the usual interaction
laws of the 4–velocity model, i.e.

$$(\mathbf{v}_1, \mathbf{v}_3) \longleftrightarrow (\mathbf{v}_2, \mathbf{v}_4) \quad .$$

The state of any knot at $t = n\Delta t$ depends on the state of all knots
of the plane lattice at $t = 0$, so that the whole process, which has been
so built, is Markovian.

Resorting to these ideas the probability of having a particle with velocity $\mathbf{v}_k$ at the (i,j)-knot at time $n\Delta t$ can be given explicitly

$$P(\alpha_{i,j}^n = 1) = [P(\alpha_{i-1,j}^{n-1} = 1) \cap P(\alpha_{i+1,j}^{n-1} \neq 3)]$$

$$\cup [P(\alpha_{i,j-1}^{n-1} = 2) \cap P(\alpha_{i,j+1}^{n-1} = 4)]$$

$$P(\alpha_{i,j}^n = 2) = [P(\alpha_{i-1,j}^{n-1} = 1) \cap P(\alpha_{i+1,j}^{n-1} = 3)]$$

$$\cup [P(\alpha_{i,j-1}^{n-1} = 2) \cap P(\alpha_{i,j+1}^{n-1} \neq 4)]$$

$$P(\alpha_{i,j}^n = 3) = [P(\alpha_{i-1,j}^{n-1} \neq 1) \cap P(\alpha_{i+1,j}^{n-1} = 3)]$$

$$\cup [P(\alpha_{i,j-1}^{n-1} = 2) \cap P(\alpha_{i,j+1}^{n-1} = 4)]$$ (3.25)

$$P(\alpha_{i,j}^n = 4) = [P(\alpha_{i-1,j}^{n-1} = 1) \cap P(\alpha_{i+1,j}^{n-1} = 3)]$$

$$\cup [P(\alpha_{i,j-1}^{n-1} \neq 2) \cap P(\alpha_{i,j+1}^{n-1} = 4)]$$

$$P(\alpha_{i,j}^n = 0) = 1 - \sum_{k=1}^{4} P(\alpha_{i,j}^n = k) \ .$$

It is now possible to develop recursively Eqs.(3.25) in order to obtain these expressions in terms of probabilities extended to any knot of the plane lattice and to any previous time (starting from the initial state), thus obtaining quantitative results. In fact, suitable rescaling of the said probabilities yields the numerical densities.

Consider again the 4–velocity model given by Eqs.(3.17) in the spatial homogeneous case

$$\frac{\partial \mathbf{N}}{\partial x} = \frac{\partial \mathbf{N}}{\partial y} = 0 \ ,$$ (3.26)

joined to the initial data

$$N_k(t = 0) = N_{k0} \ , \quad k = 1,\ldots,4 \ .$$ (3.27)

The initial value problem provides the following analytical positive solution for any $N_{k0} \in \mathbb{R}_+$

$$N_k(t) = N_{k0} + (-1)^{k+1} \frac{\mu_o}{\nu_o} \left(1 - e^{-\nu_o t}\right) \qquad (3.28)$$

where $\mu_o = N_{20}N_{40} - N_{10}N_{30}$ and $\nu_o = \sum_{k=1}^{4} N_{k0}$.

This solution tends asymptotically to the absolute Maxwellian state

$$\widehat{N}_{k\infty} = N_{k0} + (-1)^{k+1} \frac{\mu_o}{\nu_o} . \qquad (3.29)$$

Consider, in particular, the following initial conditions

$$P(\alpha_{i,j}^o = k) = P_k(t = 0) = N_{ko}$$

$$P(\alpha_{i,j}^o = 0) = P_o(t = 0) = 1 - \nu_o .$$

The results of the experiment is presented in Fig. 3.4, where the probabilities $P_1(t), \ldots, P_4(t)$ are plotted versus time. The curves are compared with the analytical results $N_1(t), \ldots, N_4(t)$ given by Eqs.(3.29).

The agreement is generally good starting also far away from the asymptotic equilibrium $N_{k\infty}$, as shown in the above figure.

As already mentioned, further developments of such a simulation scheme are necessary in order to test its validity as an alternative to discrete kinetic theory. In particular it would be certainly of interest to develop the scheme for more general physical situations: models with a large number of velocities, for gas mixtures or for chemically reacting gases.

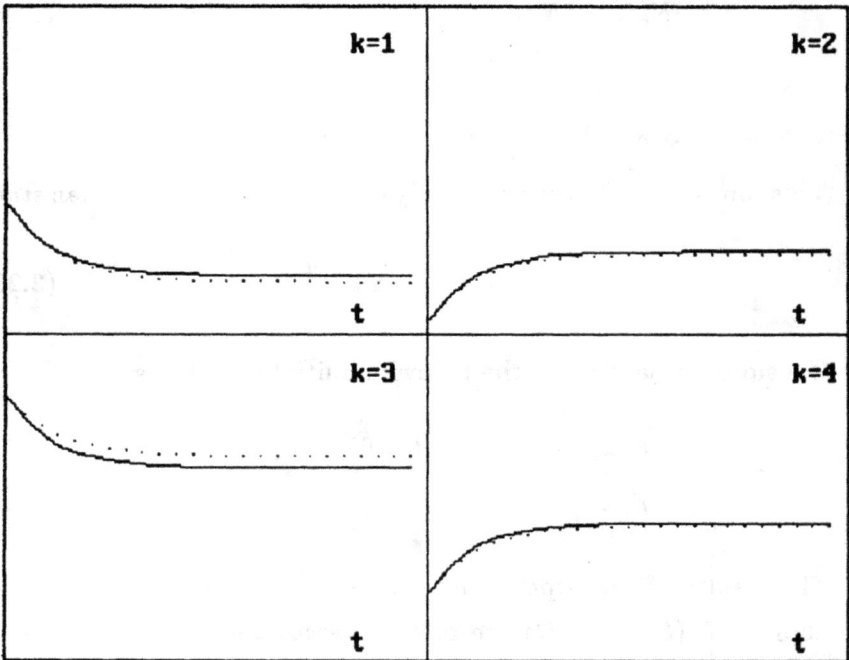

Fig. 3.4 — *Evolution of the $P_k(t)$ (dotted lines) and $N_k(t)$ (solid lines),
$k = 1, \ldots, 4$, for the 4–velocity model starting from
$N_{10} = 0.1$, $N_{20} = 0.01$, $N_{30} = 0.15$, $N_{40} = 0.05$*

References

[1] T. Platkowski and R. Illner, "Discrete velocity models of the Boltzmann equations: A survey of the mathematical aspects of the theory", *SIAM Review*, **30**, 1988, p.213.

[2] N. Bellomo and T. Gustafsson, "The discrete Boltzmann equation: A review of the mathematical aspects of the initial and initial-boundary value problems", *Review Math. Phys.*, to appear in 1991.

[3] M. Lachowicz and R. Monaco, "Existence and quantitative analysis of the solutions to the initial value problem for the discrete Boltzmann equation in all space", *SIAM J. Appl. Math.*, **49**, 1989, p.231.

[4] G. Kantorovic and L. Akilov, **Analyse Fonctionelle**, Mir, Moscow, 1977.

[5] N. Bellomo and G. Toscani, "On the Cauchy problem for the discrete Boltzmann equation with multiple collisions: Existence, uni-

queness and stability", *Stability and Appl. Anal. in Continuous Media*, **1**, to appear in 1991.

[6] N. Bellomo, A. Palczewski and G. Toscani, **Mathematical Topics in Nonlinear Kinetic Theory**, World Scientific, London, Singapore, 1988.

[7] R. DiPerna and P.L. Lions, "On the Cauchy problem for the Boltzmann equation: Global existence and weak stability", *Annals of Math.*, **130**, 1989, p.321.

[8] G. Toscani, "On the discrete velocity models of the Boltzmann equation in several dimensions", *Annali Mat. Pura Appl.*, **4**, 1984, p.297.

[9] G. Toscani, "Global existence and asymptotic behaviour for the discrete velocity models of the Boltzmann equation", *J. of Math. Phys.*, **26**, 1986, p.2918.

[10] J.M. Bony, "Existence globale a données de Cauchy petites pour le modèles discrétes de l'equation de Boltzmann", Reprint 1990.

[11] S. Kawashima, "Global existence and stability of solutions for discrete velocity models of the Boltzmann equation", *Lect. Notes in Num. Appl. Anal.*, **6**, 1983, p.58.

[12] N. Bellomo and S. Kawashima, "The discrete Boltzmann equation with multiple collisions: Global existence and stability for the initial value problem", *J. Math. Phys.*, **31**, 1990, p.245.

[13] T. Nishida and M. Mimura, "On the Broadwell's model for a simple discrete velocity gas", *Proc. Japan Acad.*, **50**, 1974, p.812.

[14] H. Cabannes, "Solution globale du problème de Cauchy on théorie cinétique discrète", *J. de Mecanique*, **17**, 1978, p.1.

[15] H. Cabannes, "Global solution of the discrete Boltzmann equation", in **Mathematical Problems in the Kinetic Theory of Gases**, Eds. H. Neunzert and D. Pack, Lang Publ. Co, Frankfurt, 1980, p.25.

[16] L. Tartar, "Existence globale pur un système hyperbolique semiliné-aire de la théorie cinétique des gaz", *Séminaire Goulaouic-Schwartz*, **1**, 1975.

[17] L. Tartar, "Some existence theorems for semilinear hyperbolic systems in one space variable", *MRC Technical Report*, **2164**, Univ. of Winsconsin, Madison, 1981.

[18] T. Beale, "Large-time behaviour of the Broadwell model of a discrete velocity gas", *Comm. Math. Phys.*, **102**, 1985, p.217.

[19] T. Beale, "Large-time behaviour of discrete velocity Boltzmann equations", *Comm. Math. Phys.*, **106**, 1986, p.659.

[20] J.M. Bony, "Solutions globales bornés pour les modèles discréte de l'equation de Boltzmann en dimension 1 d'espace", in **Equations aux Derivees Partielles**, Centre Math. Ecole Politec. Paris, 1987.

[21] H. Cabannes and S. Kawashima, "Le problème aux valeurs initiales en théorie cinétique discrète", *Comp. Rend. Acad. Sci. Paris, I*, **307**, 1988, p.507.

[22] G. Toscani, "On the Cauchy Problem for the discrete Boltzmann equation with initial values in $L_+^1(\mathbb{R})$", *Comm. Math. Phys.*, **121**, 1989, p.121.

[23] H. Cornille, "Exact $(1+1)$–dimensional solutions of discrete planar velocity Boltzmann models", *J. Statist. Phys.*, **48**, 1987, p.789.

[24] H. Cornille, "Exact solutions for discrete kinetic models with ternary

collisions", *J. Math. Phys.*, **29**, 1988, p.1667.

[25] H. Cabannes and D.H. Thiem, "Solutions exactes pour certains modèles discréte de l'equation de Boltzmann", *Comp. Rend. Acad. Sci. Paris, I*, **304**, 1987, p.29.

[26] H. Cabannes, "Solution globale des equation de Broadwell pour des donnèes initiales partielment negatives", *European J. Mech./B Fluids*, to appear in 1991.

[27] R. Bellman, B.G. Kashef and J. Casti, "Differential quadrature: A technique for the rapid solution of nonlinear partial differential equations", *J. of Comp. Physics*, **10**, 1972, p.40.

[28] N.S. Bachvalov, **Methodes Numeriques**, Mir, Moscow, 1981.

[29] D. Euvrard, **Résolution Numérique des Equations aux Dérivées Partielles**, Masson, Paris, 1988.

[30] E. Gabetta, "From stochastic mechanics to the discrete Boltzmann equation: The Broadwell model", *J. Math. Modelling Comp.*, **15**, 1991, p.1.

CHAPTER 4

SHOCK WAVES

The analysis of the structure of shock waves and of the existence and stability of travelling shock waves is a classical topic both in continuum fluid-dynamics and in kinetic theory of gases. In fact this kind of flow is similar to the ones arising near blunt bodies at small Knudsen numbers [1]. Therefore the study of shock waves is certainly of great interest in applied sciences.

Shock waves can also be studied in the framework of the Euler and Navier-Stokes description of fluid-dynamics. This latter mathematical model certainly provides accurate description of the flow conditions in the case of weak shock waves and small Knudsen numbers. On the other hand for strong shock waves and/or large Knudsen numbers the description provided by the Boltzmann equation appears to be more accurate if compared with experimental observations. The reader is referred to [1] for a wide bibliography on this topic.

One can then state that the quantitative and qualitative analysis of shock waves is certainly an interesting field of application of the discrete Boltzmann equation. In fact, as shown in ref.[2], this model provides an accurate description, both at a qualitative and quantitative level, of several interesting features of the gas dynamic flow.

Several contributions to this topic can be found in the specialized literature. Most of the papers deal either with the six–velocity Broadwell model, both for a one–component gas and for binary gas mixtures, or with the planar six–velocity model with triple collisions. Even if these models can be found in Chapter 2, they will be reported here, for sake of completness, in a dimensionless form which is suitable for the applications which are dealt with in this chapter.

In particular, this chapter provides a unified presentation of the various mathematical and physical results available in the literature. The chapter will also indicate the mathematical problems which are still open and the main difficulties to be tackled to deal with them, so that the reader can find useful indications for future research activities in the field.

The chapter is in five sections. The first one deals with the mathematical description of the problem, while the second is related to the computing of the speed of sound. The third section deals with the general methodology and with the results known in the literature on the analysis of the existence problem for travelling shock waves in the case of general discrete velocity models. This section also reviews the mathematical results on the stability of shock profiles in the case of specific models. The fourth section tackles the problem of the determination of steady profiles and of the onset of shock waves, and provides several specific applications for different discrete models. Finally the last section deals with a general discussion and with the indication of some open problems, which may be object of specific research activity in the field.

4.1 Statement of the Mathematical Problem

The general structure of the discrete Boltzmann equation is reported in this chapter for sake of completeness and for simplicity of presentation

$$\left(\frac{\partial}{\partial t} + \mathbf{v}_i \cdot \nabla_{\mathbf{x}}\right) N_i = J_i^{(2)}[\mathbf{N}] + J_i^{(3)}[\mathbf{N}] \ , \quad i = 1, \dots, n \qquad (4.1)$$

$$N_i = N_i(t, \mathbf{x}) : \ [0, T] \times \mathbb{R}^d \longmapsto \mathbb{R}_+ \ , \quad d = 1, 2, 3 \qquad (4.2)$$

where $\mathbf{v}_i$, $(i = 1, \dots, n)$ denote the admissible velocities and $J_i^{(2)}$ and $J_i^{(3)}$ are the collision operators for the binary and triple collisions respectively.

The operators have been defined in Eqs.(1.2,1.3) and are related to specific models. We recall that when the number densities are equilibrium Maxwellian densities, which will be denoted as usual by $\widehat{N}_i$, then the collision operators are equal to zero

$$J_i^{(2)}[\widehat{\mathbf{N}}] = J_i^{(3)}[\widehat{\mathbf{N}}] = 0 \quad i = 1, \dots, n \ . \qquad (4.3)$$

In spite of the relevance of the problem, dealt with in this chapter, mathematical results are available, as we shall see, only for some special models: the Broadwell model for both a simple gas and a binary gas mixture, Cabannes' 14–velocity model and the planar 6–velocity models with triple collisions. Certainly, results for general models of the discrete Boltzmann equation would be desired, however several mathematical difficulties are still waiting to be solved.

Having in mind Eqs.(4.1)–(4.3), we have all tools to provide the mathematical descriptions of the shock wave problems which will be dealt with and reviewed in this chapter :

— Existence of shock profiles,

— Onset and formation of shock waves.

Moreover a brief survey on the problem of stability of shock waves will be provided.

Since the shock wave problems studied in this chapter are all in one space dimension, in what follows we will consider

$$x \in \mathbf{R} \quad \text{and} \quad N_i = N_i(t, x) .$$

The first problem, i.e. existence of shock profiles, consists in proving the existence of shock solutions

$$N_i(x - \beta t) \tag{4.4}$$

travelling with constant velocity β and linking, asymptotically in space, two constant Maxwellian distributions

$$\widehat{N}_i^- = N_i(x \to -\infty) \quad , \quad \widehat{N}_i^+ = N_i(x \to +\infty) . \tag{4.5}$$

The two Maxwellians are not independent and need to be expressed one in term of the other, by the conservation equations. The relations among macroscopic observables characterizing the Maxwellians are called Rankine-Hugoniot conditions in analogy with those of the continuous kinetic theory [1]. Such relations, obtained at a macroscopic level, can also be transferred at the microscopic level, so that it is possible to write

$$\widehat{N}_i^+ = \widehat{N}_i^+(\widehat{N}_i^-) \ . \tag{4.6}$$

Various details of this problem have been studied by several authors. Following the pioneer work by Broadwell [3], Gatignol [4] and Cabannes [5] have studied several aspects related to the formulation of the Rankine-Hugoniot relations for a simple monoatomic gas. Gatignol has also provided analytical solutions generalizing the original results by Broadwell. Caflisch [6], starting from Gatignol's paper [4], has analysed the existence of shock profiles for the Broadwell model and the gap between these profiles and the ones described by the Navier-Stokes equation. Papers [2,7,8] deal with the computation of shock profiles for a binary gas mixture as well as with suitable comparisons [2] with profiles obtained experimentally by other authors.

Methodological aspects are dealt with in papers [9–13] which mainly refer to mathematical aspects related to the analysis of existence of shock profiles for general models of the discrete Boltzmann equation either with binary collisions only or with both binary and triple collisions.

The analysis of the stability of shock profiles essentially consists in the study of the qualitative behaviour of solutions referred to the perturbation of a travelling wave

$$N_i(x - \beta t) + \varphi_i(x) \tag{4.7}$$

where $N_i(x - \beta t)$ is a steady shock profile and φ_i a small perturbation. The only two papers where it is possible to find original contributions to this topic are due to Kawashima and Matzumura [14] and to Caflisch and Tai Ping Liu [15]. Both papers deal with the six-velocity Broadwell model for a simple gas.

The analysis of the second topic, i.e. onset and formation of shock waves, consists in the analysis of the initial value problem for Eq.(4.1) with initial conditions

$$N_{io} = \begin{cases} \widehat{N}_i^- & , \quad \text{if } x \le 0 \\ \\ \widehat{N}_i^+ & , \quad \text{if } x > 0 \end{cases} \tag{4.8}$$

where $\widehat{N}_i^-$ and $\widehat{N}_i^+$ are the Maxwellian densities defined in (4.5). Then one has to verify whether for $t \to \infty$ the solution tends to the steady solution (4.4), which is consistent with the Rankine-Hugoniot conditions.

4.2   Determination of the Velocity of Sound

The velocity of propagation of sound waves can be classically determined from the conservation equations referred to the Maxwellian state (i.e. the Euler equations) which, following Chapter 1, are given by

$$< \boldsymbol{\Phi}^{(\chi)}, \frac{\partial \widehat{\mathbf{N}}(\widehat{\mathbf{W}})}{\partial t} + [\mathbf{V}]\widehat{\mathbf{N}}(\widehat{\mathbf{W}}) >= 0 \ , \quad \chi = 1, \ldots, \delta \ , \tag{4.9}$$

where $\{\mathbf{\Phi}^{(\chi)}\}$ is an orthogonal basis of the space of collisional invariants. We recall that the hydrodynamic moments at equilibrium and the Maxwellians are respectively given by

$$\widehat{\mathbf{W}} = \{\widehat{W}_1, \ldots, \widehat{W}_\delta\} \in \mathbb{R}^\delta \quad \text{with} \quad \widehat{W}_\chi = <\mathbf{\Phi}^{(\chi)}, \widehat{\mathbf{N}}>$$

and

$$\widehat{\mathbf{N}} = \exp\left[\sum_{\chi=1}^{\delta} c_\chi \mathbf{\Phi}^{(\chi)}\right].$$

Equations (4.9) can be transformed into the equivalent set of homogeneous equations

$$\left[\mathbf{A}^{(1)}(\widehat{W}_\chi)\right] \frac{\partial}{\partial t} \begin{pmatrix} c_1 \\ \vdots \\ c_\delta \end{pmatrix} + c\left[\mathbf{A}^{(2)}(\widehat{W}_\chi)\right] \nabla_\mathbf{x} \begin{pmatrix} c_1 \\ \vdots \\ c_\delta \end{pmatrix} = \begin{pmatrix} 0 \\ \vdots \\ 0 \end{pmatrix} \quad (4.10)$$

where $[\mathbf{A}^{(1)}]$ and $[\mathbf{A}^{(2)}]$ are square matrices of rank δ whose elements are expressed in terms of the hydrodynamic moments calculated in the equilibrium state $\widehat{W}_\chi$.

A sound wave (see [9,16]) is a one-dimensional travelling solution of Eq.(4.10); the sound speed is the real characteristic velocity of the solution in one space dimension. In fact putting

$$z = x - \lambda c t ,$$

Eq.(4.10) is transformed into

$$c\left[\lambda\mathbf{A}^{(1)}(\widehat{W}_\chi) - \mathbf{A}^{(2)}(\widehat{W}_\chi)\right]\frac{\partial}{\partial z}\begin{pmatrix}c_1\\\vdots\\c_{\delta'}\end{pmatrix} = \begin{pmatrix}0\\\vdots\\0\end{pmatrix} \quad,\quad \chi = 1,\ldots\delta' \ ,$$

(4.11)

where now the matrices have rank $\delta' < \delta$, which corresponds to the dimension of the space of collisional invariants for the discrete model in one space dimension. Eq.(4.11) then admits the solution

$$\det\left[\lambda\mathbf{A}^{(1)}(\widehat{W}_\chi) - \mathbf{A}^{(2)}(\widehat{W}_\chi)\right] = 0 \qquad (4.12)$$

and the sound speed is given by one of the roots $c\lambda_\chi$, $\chi = 1,\ldots,\delta'$. In the applications proposed in this chapter we will see the physical interpretation of all the roots of Eq.(4.12).

In order to give some further insight into the calculations which have been presented above, the expression of sound speed will be derived here for the model for gas mixtures with fourteen velocity directions which has been described in Section 2.2.3. The absolute Maxwellian state is given by

$$\forall p = 1,\ldots,P \ , \quad \forall i = 1,\ldots,6 \ , \quad \forall(jk\ell) = 1,\ldots,8$$

$$\widehat{N}_i^p = \frac{\widehat{\nu}_p}{4}\left(1 - \frac{k_B}{\mu_p^2 m_p c^2}\,\widehat{T}_p\right) = N_o^p$$

$$\widehat{M}_{jk\ell}^p = \frac{\widehat{\nu}_p}{16}\left(\frac{3k_B}{\mu_p^2 m_p c^2}\,\widehat{T}_p - 1\right) = M_o^p \ ,$$

where N_i^P are the densities related to the velocities from the center of the cube to the center of the faces and $M_{jk\ell}^P$ those related to the velocities pointing to the vertices (see Eqs.(2.38)). The matrix

$$\left[\lambda \mathbf{A}^{(1)} - \mathbf{A}^{(2)}\right]$$

has rank $\delta' = 1 + 2P$ (conservation of mass and energy for each species, and of the x-component of the momentum for the overall mixture). The structure of the above matrix has been derived in [16] and results to be diagonal with elements

$$\left[\mathbf{A}^{(1)}\right] = \text{diag}\{A_i^{(1)}\} \quad , \quad \left[\mathbf{A}^{(2)}\right] = \text{diag}\{A_i^{(2)}\}$$

$$i = 1, 2, 3: \quad A_i^{(1)} = \frac{1}{5P} \left(\sum_{p=1}^{P} N_o^p + 4 \sum_{p=1}^{P} M_o^p\right) \quad , \quad A_i^{(2)} = 0$$

$$\text{even } i \geq 4: \quad A_i^{(1)} = \sum_{p=1}^{P} N_o^p \quad , \quad A_i^{(2)} = \frac{1}{\sqrt{15P}} \sum_{p=1}^{P} \mu_p N_o^p$$

$$\text{odd } i \geq 5: \quad A_i^{(1)} = \sum_{p=1}^{P} M_o^p \quad , \quad A_i^{(2)} = \sqrt{\frac{4}{5P}} \sum_{p=1}^{P} \mu_p M_o^p \ .$$

The solution of Eq.(4.12) is then

$$\lambda^{2P-1} \left(\lambda^2 - \frac{\displaystyle\sum_{p=1}^{P} \mu_p^2 N_o^p + 12 \sum_{p=1}^{P} \mu_p^2 M_o^p}{\displaystyle 3 \sum_{p=1}^{P} N_o^p + 12 \sum_{p=1}^{P} M_o^p}\right) = 0 \ , \qquad (4.13)$$

so that one finds $2P - 1$ roots equal to zero and two roots

$$\lambda = \pm \sqrt{\frac{\displaystyle\sum_{p=1}^{P} \mu_p^2 N_o^P + 12 \sum_{p=1}^{P} \mu_p^2 M_o^P}{3 \displaystyle\sum_{p=1}^{P} N_o^P + 12 \sum_{p=1}^{P} M_o^P}} \tag{4.14}$$

which correspond to the sound speed $c\lambda$ in the two directions of the x–axis.

It is important to note that the result (4.14) also provides the particular value of sound speed for Cabannes 14–velocity model when $P = 1$. Moreover also the sound speed of the Broadwell model with P gas components can be recovered setting $\forall p : \quad M_o^P = 0$. This result will be used, as we shall see, in Section 4.4.

4.3   Existence and Stability of Shock Waves

The problem of existence and stability of shock waves in discrete kinetic theory is certainly a challenging (and difficult) topic for applied mathematicians. In spite of the fact that this matter has been studied for several years and in spite of several interesting and carefully obtained numerical experiments which show existence and stability of shock waves, very few theoretical results have been achieved. The known results are almost completely referred to the Broadwell model for a simple gas.

This section will provide a description of the methodological line which needs to be followed in order to provide mathematical proofs for such a problem. The analysis will be first presented for a general discrete velocity model for a simple gas with a velocity discretization with only one velocity modulus. This restriction will simplify several notations and will have the advantage of relaying upon some mathematical results given, for such a class of models, by Kawashima and Bellomo [10]. Once the analysis of the problem for a simple gas has been developed, some brief indications will be given to deal with gas mixtures and, more in general, with models with more velocity moduli.

Consider then Eq.(4.1) in one-space dimension for a regular model with only one velocity modulus and for a one-component gas. The regularity of the model assures that the dimension of the space $\mathcal{M}$ is equal to two (corresponding to conservation of mass and momentum along the x-axis). Therefore the independent macroscopic observables are the mass density ρ and the x-component u of the mean velocity.

Hence the Maxwellian state is uniquely defined in terms of two parameters and, particularizing Eq.(1.49) to this case, can be written as

$$\widehat{N}_i = A e^{c_x v_{ix}} \ , \ A \in \mathbb{R}_+ \ , \ c_x \in \mathbb{R} \ , \ i = 1, \ldots, n \qquad (4.15)$$

which is the case of several well structured models such as the planar 6–velocity model with binary and triple collisions studied in Section 2.1.4 or the spatial six–velocity Broadwell model presented in Section 2.2.1. The analysis of the problem will be developed in five steps.

Step 1: Following the procedure outlined at the end of Section 1.4.1, the original system (4.1) of n differential equation, is written in terms of two conservation equations and $n - 2$ equations written in non-conservative form

$$\frac{\partial \rho}{\partial t} + \frac{\partial (\rho u)}{\partial x} = 0$$

$$\frac{\partial (\rho u)}{\partial t} + \frac{\partial (\rho \sigma)}{\partial x} = 0 \qquad (4.16)$$

$$\frac{\partial f_k}{\partial t} + \frac{\partial g_k}{\partial x} = h_k \quad , \quad k = 1, \ldots, n - 2$$

where ρ is the mass density, ρu is the x-component of the momentum, while σ is

$$\sigma = \sum_{i=1}^{n} N_i v_{ix}^2$$

and $f_k = f_k[\mathbf{N}]$, $g_k = g_k[\mathbf{N}]$, $h_k = h_k[\mathbf{N}]$ are suitable functions of the densities N_i (to be determined for each model).

Note that the first two equations in (4.16) do not represent a closed system. In fact σ is not, in general, a function of ρ and u. Therefore one cannot solve the first two equations independently from the remaining ones.

Step 2: The second step consists in the analysis of the properties of the Euler equations given in the previous section. These can be re-written in this case as

$$\frac{\partial}{\partial t} \begin{pmatrix} \widehat{\rho} \\ \widehat{u} \end{pmatrix} + (\mathbf{A}(\widehat{\rho}, \widehat{u}, \sigma)) \frac{\partial}{\partial x} \begin{pmatrix} \widehat{\rho} \\ \widehat{u} \end{pmatrix} = \begin{pmatrix} 0 \\ 0 \end{pmatrix} , \qquad (4.17)$$

where A, c_x are expressed in terms of $\widehat{\rho}, \widehat{u}$. Moreover for the Maxwellian state given by Eq.(4.15), σ can be regarded as a function of $\widehat{u}$ only, so that one has

$$\hat{\rho} = \hat{\rho}(A, c_x) , \quad \hat{u} = \hat{u}(c_x) , \quad \sigma = \sigma(\hat{u}) , \tag{4.18}$$

and the matrix $[\mathbf{A}]$ is defined as

$$[\mathbf{A}(\hat{\rho}, \hat{u}, \sigma)] = \begin{pmatrix} \hat{u} & \hat{\rho} \\ [\sigma(\hat{u}) - \hat{u}^2]/\hat{\rho} & \frac{d\sigma}{du}(\hat{u}) - \hat{u} \end{pmatrix} . \tag{4.19}$$

As known [10], it is important to analyse the properties of the matrix $[\mathbf{A}]$ in order to prove genuine nonlinearity

$$\forall \hat{\rho}, \hat{u} , \quad \nabla \lambda_j(\hat{u}) \cdot \mathbf{r}_j(\hat{\rho}, \hat{u}) \neq 0 , \quad j = 1, 2 , \tag{4.20}$$

and strict hyperbolicity

$$\lambda_1(\hat{u}) < \hat{u} < \lambda_2(\hat{u}) , \quad 2\lambda_1(\hat{u}) < \frac{d\sigma}{du}(\hat{u}) < 2\lambda_2(\hat{u}) \tag{4.21}$$

where λ_j and $\mathbf{r}_j$ are, respectively, the eigenvalues and eigenvectors of the matrix $[\mathbf{A}]$.

Step 3: The third step consists in a further analysis of the Euler equations. In fact one has to prove suitable compatibility conditions on the shock wave problem for the discrete equation, so that necessary and sufficient conditions hold for the existence of shock waves described by the Euler equations.

First, one applies the change of variables $z = x - \beta t$ to the first two equations in (4.16) in order to obtain a system of ordinary differential equations

$$\frac{d}{dz}(-\beta\rho + \rho u) = 0$$

$$\frac{d}{dz}(-\beta\rho u + \rho\sigma) = 0 \ , \tag{4.22}$$

then an integration between the Maxwellian states at $z \to \pm\infty$ yields the Rankine-Hugoniot relations

$$-\beta(\widehat{\rho}_+ - \widehat{\rho}_-) + (\widehat{\rho}_+\widehat{u}_+ - \widehat{\rho}_-\widehat{u}_-) = 0$$

$$-\beta(\widehat{\rho}_+\widehat{u}_+ - \widehat{\rho}_-\widehat{u}_-) + [\widehat{\rho}_+\sigma(\widehat{u}_+) - \widehat{\rho}_-\sigma(\widehat{u}_-)] = 0 \ . \tag{4.23}$$

It is well known, see [1,10,11] and the bibliography therein included, that the weak solution of the Euler equations are given by

$$(\widehat{\rho}, \widehat{u}) = \begin{cases} (\widehat{\rho}_-, \widehat{u}_-) \ , & \text{if } x < \beta t \\ \\ (\widehat{\rho}_+, \widehat{u}_+) \ , & \text{if } x > \beta t \end{cases} \tag{4.24}$$

where the upstream conditions $(\widehat{\rho}_-, \widehat{u}_-)$ are defined by the statement of the problem and the downstream conditions $(\widehat{\rho}_+, \widehat{u}_+)$ are the one delivered by Eqs.(4.23) at fixed values of the propagation velocity β.

It is therefore necessary to show existence and uniqueness of the solution of (4.22) for fixed values of $(\widehat{\rho}_-, \widehat{u}_-, \beta)$. This solution is defined *global* if it is obtained for all admissible values of β, namely for $|\beta|$ spanning from 0, which corresponds to $(\widehat{\rho}_-, \widehat{u}_-) = (\widehat{\rho}_+, \widehat{u}_+)$, to the maximum admissible value of $|\beta|$, which correspond to $\widehat{u}_+ = 0$.

Examples of global solution of the Euler equations, for discrete velocity models of the Boltzmann equation, can be found in [10,11].

Moreover, the aforementioned solution can be regarded physically reasonable if

$$\lambda_1(\widehat{u}_+) < \beta < \lambda_1(\widehat{u}_-) \quad \text{and} \quad \beta < \lambda_2(\widehat{u}_+) \tag{4.25}$$

$$\lambda_2(\widehat{u}_+) < \beta < \lambda_2(\widehat{u}_-) \quad \text{and} \quad \beta > \lambda_1(\widehat{u}_-) . \tag{4.26}$$

The discontinuity at $z = 0$ is then called *shock wave*.

The conditions defined above (which need to be verified for each particular model of the discrete Boltzmann equation) can be regarded as sufficient conditions for the existence of shock waves according to the Euler description.

Step 4: This step consists in the analysis of existence of shock wave solutions of Eqs.(4.16) once the analysis developed in Step 3 has been completed. In fact the Rankine-Hugoniot equations can be used to obtain a system of $(n - 2)$ equations in $(n - 2)$ unknowns. This system can be put in terms of ordinary differential equations by means of the change of variable $z = x - \beta t$.

Existence of solutions can be obtained by the qualitative analysis of the system of differential equations which has been obtained in such a way.

The reader can practically deal with this step referring to the Broadwell model defined in Section 2.2.1. Then one should obtain two conservation equations which can be easily integrated generating only one equation written in non-conservative form (as known [15], the solution of this equation can be obtained in a closed analytical form).

Step 5: The fifth step consists in the qualitative analysis of the solutions for *small* perturbations (see Eq.(4.7)) of the travelling shock wave. The shock solution is stable if one can prove that, for sufficiently small perturbations, the solution decays asymptotically in time to the travelling wave solution.

Referring to the program which has been described above, it needs to be mentioned that mathematical results are known only for the Broadwell model for a simple gas. In fact for such a model the analysis of the first three steps is a very simple problem, see [4] and [6]. Moreover an analytic solution can be obtained with reference to Step 4.

These solutions [6] become the pertinent basis for the development of the stability analysis described in Step 5 as shown in papers [14,15].

The same program does not seem easy to be completed for more general models of the discrete Boltzmann equation and in fact it is not in the pertinent literature. The crucial difficulty is the lack of general proofs referred to Step 4. Once such proofs were available, the stability result by Kawashima and Matsumura [14] would certainly provide stability indications.

On the other hand the relatively simpler aims of Steps 1–3 have been solved for regular discrete velocity models. It should be possible the generalization of the result of [10] to wider classes of models, i.e. models characterized by more than one velocity moduli or models for gas mixtures. In this last case the number of conservation equations will be related to the number of species in the mixture.

The application dealt with in the next section will show the practical handling of some of the calculations described in this present section.

4.4 Propagation and Structure of Shock Waves

Applications of shock wave problems are here considered for some of the models introduced in Chapter 2. In details we will deal with shock waves provided by the Broadwell model for a binary gas mixture and by the planar six-velocity model with triple collisions. Moreover we will be concerned with the shock structure referred to the steady shock wave problem formulated by Eqs.(4.4–4.6) and with the shock wave onset problem formulated by Eqs.(4.1) joined to the initial data (4.8). Then some quantitative results will be provided for the aforementioned two models.

4.4.1 *Shock Waves by the 2×6–Broadwell Model for a Binary Gas Mixture*

Considering that both this and the next section deal with one-dimensional applications, it is useful to introduce the following dimensionless quantities

$$\tau = \frac{ct}{\ell} \ , \quad \eta = \frac{x}{\ell} \ , \quad \mathbf{f} = \frac{\mathbf{N}}{\mathcal{N}} \ , \tag{4.27}$$

where ℓ and $\mathcal{N}$ are respectively a reference length and a reference density to be specified.

Referring to the notations already used in Chapter 2, recall that the Broadwell model for a binary mixture of gases with molecular masses m_1 and m_2, respectively, has the following velocity discretization

$$\mathbf{v}_i^1 = \begin{cases} c\mathbf{e}_i \ , & i=1,2,3 \\ \\ -c\mathbf{e}_{i-3} \ , & i=4,5,6 \ , \end{cases} \qquad (4.28a)$$

for the first gas and

$$\mathbf{v}_i^2 = \begin{cases} \mu c\mathbf{e}_i \ , & i=1,2,3 \\ \\ -\mu c\mathbf{e}_{i-3} \ , & i=4,5,6 \ , \end{cases} \qquad (4.28b)$$

for the second gas, where $\mathbf{e}_1$, $\mathbf{e}_2$ and $\mathbf{e}_3$ are the unit vectors of an orthogonal frame and $\mu = m_1/m_2$. In one space dimension the model equations (2.32) provide only six independent number densities, since

$$p = 1,2 \ : \ N_2^p = N_3^p = N_5^p = N_6^p \ ,$$

so that, in order to describe the behaviour of the gas mixture, one can study the evolution of the following normalized number densities

$$f_1 = \frac{N_1^1}{\mathcal{N}} \ , \quad f_2 = \frac{N_2^1}{\mathcal{N}} \ , \quad f_3 = \frac{N_4^1}{\mathcal{N}} \ ,$$
$$f_4 = \frac{N_1^2}{\mathcal{N}} \ , \quad f_5 = \frac{N_2^2}{\mathcal{N}} \ , \quad f_6 = \frac{N_4^2}{\mathcal{N}} \ . \qquad (4.29)$$

If S_{11}, S_{22} and S_{12} are the cross sectional areas for the collisions between particles of the first gas, of the second one and of the pair respectively, and if the reference length ℓ in (4.27) is chosen to be

$$\ell = \frac{1}{S_{11}\mathcal{N}} \, ,$$

then detailed calculations lead to the following one dimensional model

$$\frac{\partial f_1}{\partial \tau} + \frac{\partial f_1}{\partial \eta} = F_1(\mathbf{f})$$

$$\frac{\partial f_2}{\partial \tau} = F_2(\mathbf{f})$$

$$\frac{\partial f_3}{\partial \tau} - \frac{\partial f_3}{\partial \eta} = F_3(\mathbf{f})$$

$$\frac{\partial f_4}{\partial \tau} + \mu\frac{\partial f_4}{\partial \eta} = F_4(\mathbf{f}) \qquad (4.30)$$

$$\frac{\partial f_5}{\partial \tau} = F_5(\mathbf{f})$$

$$\frac{\partial f_6}{\partial \tau} - \mu\frac{\partial f_6}{\partial \eta} = F_6(\mathbf{f})$$

where

$$\tau = \frac{ct}{S_{11}\mathcal{N}} \, , \quad \eta = \frac{x}{S_{11}\mathcal{N}}$$

and

$$F_1(\mathbf{f}) = \frac{4}{3}(f_2^2 - f_1 f_3) + 2\mathcal{K}_1(2f_2 f_5 + f_4 f_3 - 3f_1 f_6)$$
$$+ 4\mathcal{K}_2(f_4 f_2 - f_1 f_5)$$

$$F_2(\mathbf{f}) = \frac{2}{3}(f_1 f_3 - f_2^2) + \mathcal{K}_1(f_1 f_6 + f_4 f_3 - 2f_2 f_5)$$
$$+ \mathcal{K}_2[f_5(f_1 + f_3) - f_2(f_4 + f_6)]$$

$$F_3(\mathbf{f}) = \frac{4}{3}(f_2^2 - f_1 f_3) + 2\mathcal{K}_1(2f_2 f_5 + f_1 f_6 - 3f_4 f_3)$$
$$+ 4\mathcal{K}_2(f_6 f_2 - f_3 f_5)$$

$$F_4(\mathbf{f}) = 2\mathcal{K}_3(f_5^2 - f_4 f_6) + 2\mathcal{K}_1(2f_2 f_5 + f_1 f_6 - 3f_4 f_3) \qquad (4.31)$$
$$+ 4\mathcal{K}_2(f_1 f_5 - f_4 f_2)$$

$$F_5(\mathbf{f}) = \mathcal{K}_3(f_4 f_6 - f_5^2) + \mathcal{K}_1(f_4 f_3 + f_1 f_6 - 2f_2 f_5)$$
$$+ \mathcal{K}_2[f_2(f_4 + f_6) - f_5(f_1 + f_3)]$$

$$F_6(\mathbf{f}) = 2\mathcal{K}_3(f_5^2 - f_4 f_6) + 2\mathcal{K}_1(2f_2 f_5 + f_4 f_3 - 3f_1 f_6)$$
$$+ 4\mathcal{K}_2(f_5 f_3 - f_2 f_6) \ ,$$

$$\mathcal{K}_1 = \frac{1+\mu}{6}\frac{S_{12}}{S_{11}} \ , \quad \mathcal{K}_2 = \sqrt{1+\mu^2}\,\frac{S_{12}}{S_{11}} \ , \quad \mathcal{K}_3 = \frac{2}{3}\mu\frac{S_{22}}{S_{11}} \ .$$

Letting $f_4 = f_5 = f_6 = 0$ leads to the celebrated Broadwell model for a simple gas, which in one space dimension writes

$$\frac{\partial f_1}{\partial \tau} + \frac{\partial f_1}{\partial \eta} = \frac{4}{3}(f_2^2 - f_1 f_3)$$

$$\frac{\partial f_2}{\partial \tau} = \frac{2}{3}(f_1 f_3 - f_2^2) \qquad (4.32)$$

$$\frac{\partial f_3}{\partial \tau} - \frac{\partial f_3}{\partial \eta} = \frac{4}{3}(f_2^2 - f_1 f_3) ,$$

where now $\tau = \frac{ct}{SN}$, $\eta = \frac{x}{SN}$ and S is the cross sectional area of the particles of the simple gas.

The mathematical formulation of the shock wave onset problem has been provided in Section 4.1. One has the initial value problem with initial conditions given by the two Maxwellians $\widehat{N}_i^-$ and $\widehat{N}_i^+$ defined, respectively, for $x \le 0$ and $x > 0$.

Then one solves the initial value problem (on the basis of classical methods of numerical analysis) to obtain quantitative results. The crucial point consists in observing the time behaviour of the propagation speed of the shock wave in order to verify that such a speed, say $\beta = \beta(t)$, tends, asymptotically in time for $t \to \infty$, to the value predicted by the solution of the Rankine-Hugoniot equations.

Classically, in order to apply solution techniques, one needs a suitable existence theorem for the solution of the initial value problem. The theorem should provide detailed information on the estimate of the existence time-interval and on the regularity of the solution. These information allow to apply safely numerical methods and to estimate convergence results.

A review of existence results has already been provided in Chapter 3 and therefore the reader finds there sufficient information on this topic.

In order to show an application on some aspects of the problem, we consider now Eqs.(4.30) and study the shock wave formation in one space dimension.

The analysis is developed in two steps: first we solve the Rankine-Hugoniot relations in order to find $\widehat{N}_i^+ = \widehat{N}_i^+(\widehat{N}_i^-)$, then we solve the initial value problem for Eqs.(4.30) with the initial conditions of the type (4.8). The solution should represent the formation of a shock wave travelling with the constant velocity predicted by the Rankine-Hugoniot equations.

Keeping this in mind and following the line indicated in Section 4.3, we first observe that the application of the change of variable

$$z = \eta - \beta \tau$$

and a suitable linear combination of Eqs.(4.30) leads to the following conservation equations

$$(1 - \beta)\frac{df_1}{dz} - (1 + \beta)\frac{df_3}{dz} - 4\beta\frac{df_2}{dz} = 0$$

$$(\mu - \beta)\frac{df_4}{dz} - (\mu + \beta)\frac{df_6}{dz} - 4\beta\frac{df_5}{dz} = 0 \qquad (4.33)$$

$$(1 - \beta)\frac{df_1}{dz} + (1 + \beta)\frac{df_3}{dz} + (\mu - \beta)\frac{df_4}{dz} + (\mu + \beta)\frac{df_6}{dz} = 0 \ .$$

The number of conservation equation is now three corresponding to conservation of mass for each species and conservation of the x–component of the momentum.

Assume now, according to the statement of the problem, to the symbols used to denote Maxwellians and to Eq.(2.33), that the upstream

Maxwellian state is given now by

$$\widehat{f_1^-} = A_- \, e^b \ , \quad \widehat{f_2^-} = A_- \ , \quad \widehat{f_3^-} = A_- \, e^{-b} \ ,$$
$$\widehat{f_4^-} = B_- \, e^b \ , \quad \widehat{f_5^-} = B_- \ , \quad \widehat{f_6^-} = B_- \, e^{-b} \ ,$$

$$(4.34)$$

and that the downstream conditions correspond to a Maxwellian with zero mass velocity

$$\widehat{f_1^+} = \widehat{f_2^+} = \widehat{f_3^+} = A_+ \ , \quad \widehat{f_4^+} = \widehat{f_5^+} = \widehat{f_6^+} = B_+ \ . \qquad (4.35)$$

The integration of Eqs.(4.33) between the limit boundaries $z \rightarrow -\infty$ and $z = \rightarrow +\infty$ yields an algebraic system solvable with respect to the unknown constants defined in (4.35), i.e. A_+, B_+, plus β

$$A_+ = \frac{A_-(\cosh b + 2)\beta - A_- \sinh b}{3\beta}$$

$$B_+ = \frac{B_-(\cosh b + 2)\beta - \mu B_- \sinh b}{3\beta} \ , \qquad (4.36)$$

and

$$\beta = \frac{(A_- + \mu B_-)(\cosh b - 1) - \Gamma}{3(A_- + B_-)\sinh b} \ , \qquad (4.37)$$

where

$$\Gamma = [(A_- + \mu B_-)^2(\cosh b - 1)^2 + 3(A_- + B_-)(A_- + \mu^2 B_-)\sinh^2 b]^{1/2} . \qquad (4.38)$$

The Rankine-Hugoniot equations, in this particular case, are then solved. The solution can be shown to be unique for number densities A_+

and B_+ non-negative. A simple algebraic analysis also shows that β is negative, which corresponds to a backward wave. Of course, exchanging the asymptotic limit conditions will provide a forward wave.

In the computation shown above the downstream conditions were assumed correspondent to the absolute Maxwellian state. As we have seen in Step 3 of Section 4.3, this corresponds, at fixed values of the upstream parameters A_- , B_- , b , to the maximum admissible value of $|\beta|$.

On the other hand the conservation equations (4.33), which correspond to the Rankine-Hugoniot conditions, can be globally solved for the full range of propagation velocity β. This problem is left to the reader as a proposed exercise.

Having in mind fluid-dynamic applications, it is useful to express the constants which define the upstream conditions in terms of characteristic numbers of the flow. In fact, one can compute from (4.34) the numerical densities of the two gases computed at $z \to -\infty$ as

$$\widehat{\nu}_1 = \mathcal{N} A_-(e^b + e^{-b} + 4)$$
$$\widehat{\nu}_2 = \mathcal{N} B_-(e^b + e^{-b} + 4)$$

together with the mean velocity of the overall mixture

$$\widehat{u} = c\frac{\mathcal{N}}{\widehat{\nu}_1 + \widehat{\nu}_2}(A_- + B_-)(e^b - e^{-b}) \ .$$

Then one can define the Mach number

$$Ma = \frac{\widehat{u}}{c\lambda} \tag{4.39}$$

where the speed of sound $c\lambda$, for the Broadwell model for a binary gas mixture, can be computed using Eqs.(4.14) with $P=2$

$$c\lambda = c\sqrt{\frac{\widehat{\nu}_1 + \mu^2 \widehat{\nu}_2}{3(\widehat{\nu}_1 + \widehat{\nu}_2)}} \ .$$

Simple calculations then yield

$$A_- = \frac{1}{2(\cosh b + 2)}$$

$$B_- = \frac{1 - \varphi}{2\varphi(\cosh b + 2)} \tag{4.40}$$

$$b = \cosh^{-1}\left[\frac{10 Ma^2}{9 + (\frac{100}{81} Ma^4 + \frac{25}{9} Ma^2 + 1)^{1/2}}\right] \ .$$

where

$$\varphi = \frac{\widehat{\nu}_1}{\widehat{\nu}_1 + \widehat{\nu}_2} \tag{4.41}$$

is the concentration of the first gas with respect to the whole mixture.

If now the same calculations are performed for the Broadwell model defined by Eqs.(4.32) with data

$$\widehat{f}_1^- = 1 \ , \quad \widehat{f}_2^- = \widehat{f}_3^- = 0 \ , \quad \widehat{f}_1^+ = \widehat{f}_2^+ = \widehat{f}_3^+ = A_+ = \frac{2}{3} \tag{4.42}$$

which correspond to the so-called infinite Mach number boundary conditions, the problem knows the explicit analytical solution given by Broadwell [3]

$$f_1(z) = \frac{1 + e^z}{1 + 3e^z}$$

$$f_2(z) = f_3(z) = 2[1 - f_1(z)] \tag{4.43}$$

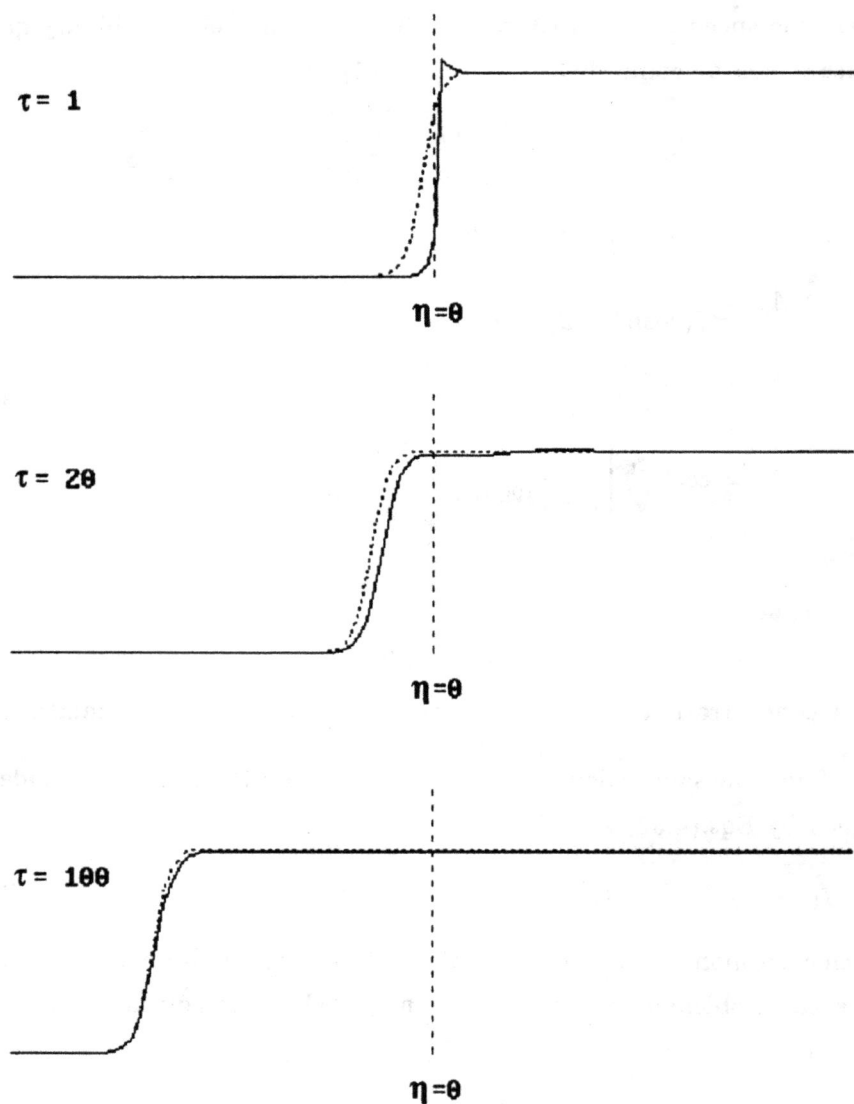

Fig. 4.1 — *Numerical density versus space (at fixed time) for a simple gas*

which describes a shock wave travelling with constant speed $\beta = -\frac{1}{3}$. This analytic solution can be used to test numerical solutions.

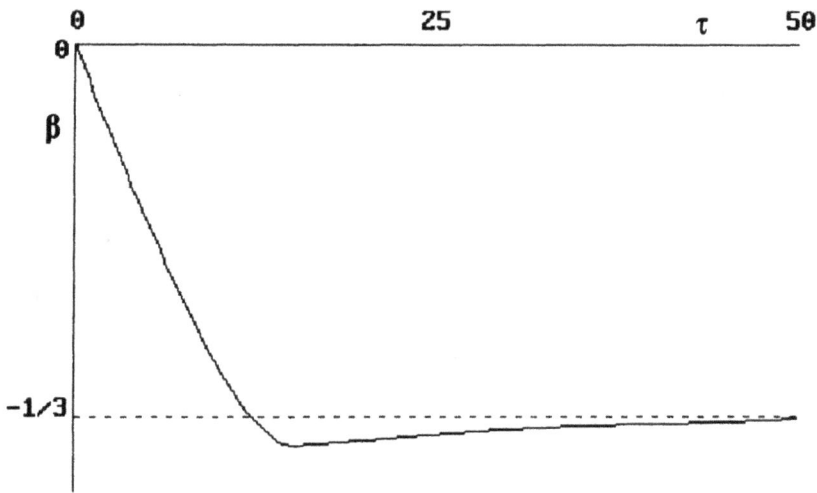

Fig. 4.2 — *Transient behaviour of the shock wave propagation speed*

The results of some practical calculations are shown in Figs. 4.1–4.4. In particular Fig. 4.1 refers to a one component gas where the dashed line is the analytic travelling wave solution (4.43) due to Broadwell. The figure clearly shows how the transient solution tends, asymptotically in time, to the travelling wave solution.

Figure 4.2 refers to the same problem and reports the time-behaviour of the propagation velocity which tends, asymptotically in time to the constant value $\beta = -\frac{1}{3}$.

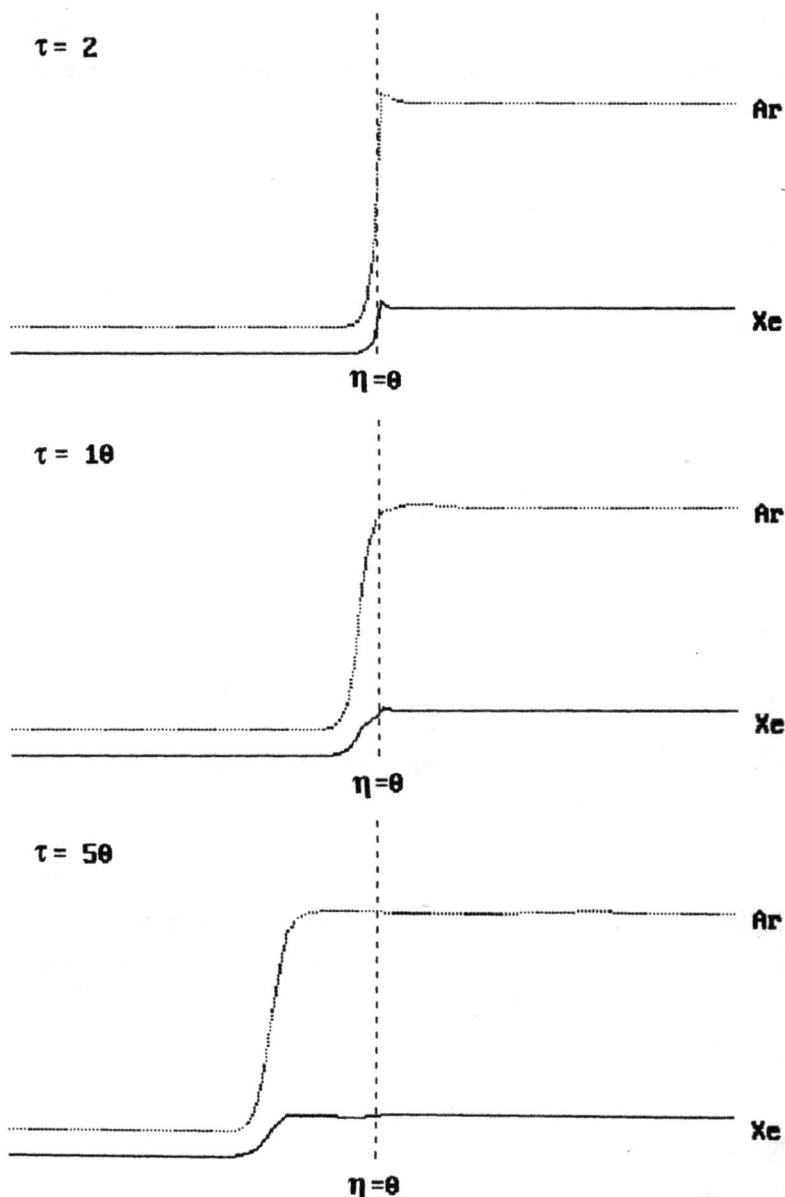

Fig. 4.3 — *Ar–Xe gas mixture: time-behaviour of the shock propagation*

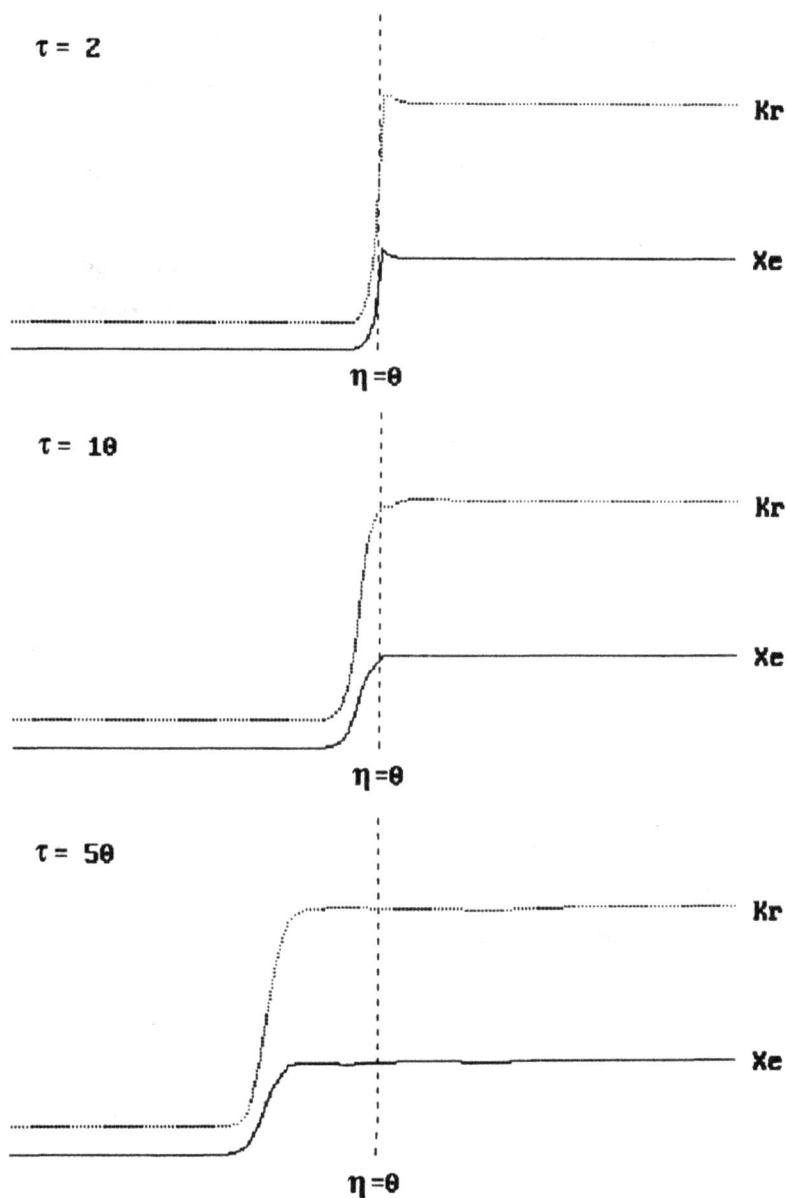

Fig. 4.4 — *Kr–Xe gas mixture: time-behaviour of the shock propagation*

Analogous results and behaviour of numerical densities are shown in Figs. 4.3 and 4.4 which refer respectively to binary gas mixtures of 60% Ar–40% Xe and 60% Kr–40% Xe for $Ma = 2$. The asymptotic values of the propagation speed are respectively $\beta = -0.2919$ and $\beta = -0.3084$.

In particular, we point out that, in the transient behaviour, the propagation velocity differs for the two gases. However it tends, asymptotically in time, to the same value predicted by the solution of the Rankine-Hugoniot relations. More details on this kind of analysis and simulation can be found in [8].

An interesting physical quantity to compute is the shock wave thickness defined, in terms of the mass density ρ, as

$$\Delta = \frac{\rho_\infty^+ - \rho_\infty^-}{\max\{\frac{d\rho}{dz}\}} \, .$$ (4.44)

which may be referred to the quantity

$$\ell = \frac{1}{S_{11}\mathcal{N}} \, .$$ (4.45)

Some results for a Helium-Xenon mixture at $Ma = 2$ are shown in Fig. 4.5 where the triangles correspond to Hamel's solution of the *linearized* continuous Boltzmann equation [17].

Analogous result is the one of Fig. 4.6 which refers to a Helium-Argon mixture. In this case the triangles refer to a solution by Sirovich [18] obtained by an analysis similar to the one of [17].

It is interesting to notice the agreement between the two solutions and the influence of the concentration of the lighter gas on the shock thickness.

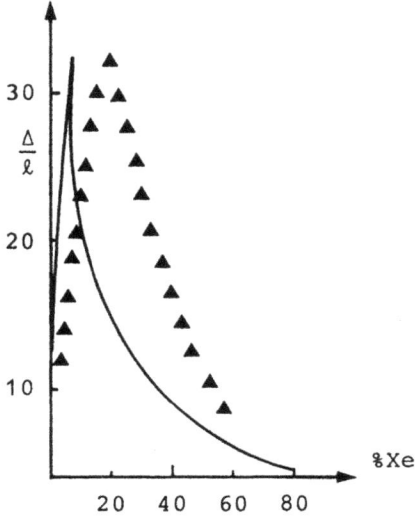

Fig. 4.5 — *Shock thickness versus Xe concentration*

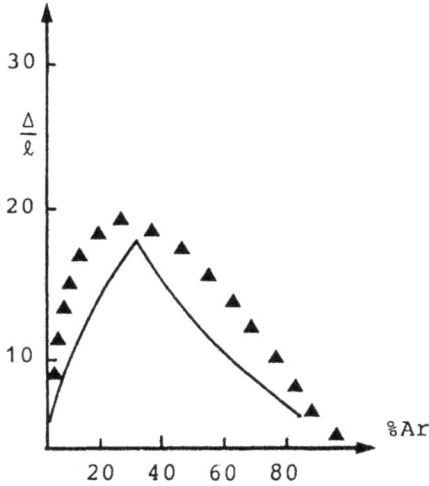

Fig. 4.6 — *Shock thickness versus Ar concentration*

4.4.2   *Shock Waves by the Planar 6–Velocity Model with Triple Collisions*

Models with triple collisions can be used to show the influence of small Knudsen numbers on the structure of shock waves. Calculations of this type were performed in [9]. Some results are here briefly reported.

Using the change of variables (4.27) with ℓ being now a fixed laboratory lenght scale, the planar 6–velocity model with triple collisions given by Eqs.(2.15) can be re-written in one space dimension as

$$\frac{\partial f_1}{\partial \tau} + \frac{\partial f_1}{\partial \eta} = 2\mathcal{K}_1(f_2 f_3 - f_1 f_4) + \mathcal{K}_2(f_2^2 f_4 - f_1 f_3^2)$$

$$\frac{\partial f_2}{\partial \tau} + \frac{1}{2}\frac{\partial f_2}{\partial \eta} = \mathcal{K}_1(f_1 f_4 - f_2 f_3) + \mathcal{K}_2(f_1 f_3^2 - f_2^2 f_4)$$

$$\frac{\partial f_3}{\partial \tau} - \frac{1}{2}\frac{\partial f_3}{\partial \eta} = \mathcal{K}_1(f_1 f_4 - f_2 f_3) + \mathcal{K}_2(f_2^2 f_4 - f_1 f_3^2) \qquad (4.46)$$

$$\frac{\partial f_4}{\partial \tau} - \frac{\partial f_4}{\partial \eta} = 2\mathcal{K}_1(f_2 f_3 - f_1 f_4) + \mathcal{K}_2(f_1 f_3^2 - f_2^2 f_4)$$

where now $f_i = \frac{N_i}{N}$, $f_5 = f_3$ and $f_6 = f_2$,

$$\mathcal{K}_1 = \frac{2}{3Kn} \quad , \quad \mathcal{K}_2 = \frac{5}{4\sqrt{\pi}}\frac{Kn_2}{Kn^2}$$

and where $Kn = \frac{1}{SN\ell}$ is a Knudsen number and $Kn_2 = \frac{\sqrt{S}}{\ell}$ is a number of the order of 10^{-5}.

The reason we used a different rescaling in this subsection is that here we want to study the effect of the density of the gas on the shock profile, while in the preceding subsection we wanted to stress the effect of the interaction between the two species on the structure of the shock profile.

In one dimension the Maxwellian state (see Section 2.1.4) is defined by

$$\widehat{\mathbf{f}} = A\{e^b, e^{b/2}, e^{-b/2}, e^{-b}\} \ . \tag{4.47}$$

Then one can join to Eqs.(4.46) the initial data

$$\mathbf{f}(0, x \le 0) = \widehat{\mathbf{f}}^-(A_-, b_-)$$
$$\mathbf{f}(0, x > 0) = \widehat{\mathbf{f}}^+(A_+, b_+) \tag{4.48}$$

where A_+ , b_+ are computed in terms of A_- , b_- from the usual Rankine-Hugoniot conditions. Then the shock onset can be studied as described in Section 4.4.1.

In particular in [9] this procedure has been followed putting $b_+ = 0$ (absolute downstream Maxwellian). Then the Rankine-Hugoniot conditions provide

$$A_+ = \frac{A_-}{6}\Gamma_1\left[\frac{1}{2} + \Gamma_2 + \sqrt{\left(\Gamma_2 - \frac{1}{2}\right)^2 + 2\Gamma_3^2}\right]$$

$$\beta = \Gamma_3\left(6\frac{A_+}{\Gamma_1} - 1\right)^{-1} \tag{4.49}$$

where

$$\Gamma_1 = A_- \left(\cosh b_- + 2 \cosh \frac{b_-}{2} \right)$$

$$\Gamma_2 = \frac{A_-}{\Gamma_1} \left(\cosh b_- + \frac{1}{2} \cosh \frac{b_-}{2} \right) \qquad (4.50)$$

$$\Gamma_3 = \frac{A_-}{\Gamma_1} \left(\sinh b_- + \sinh \frac{b_-}{2} \right) .$$

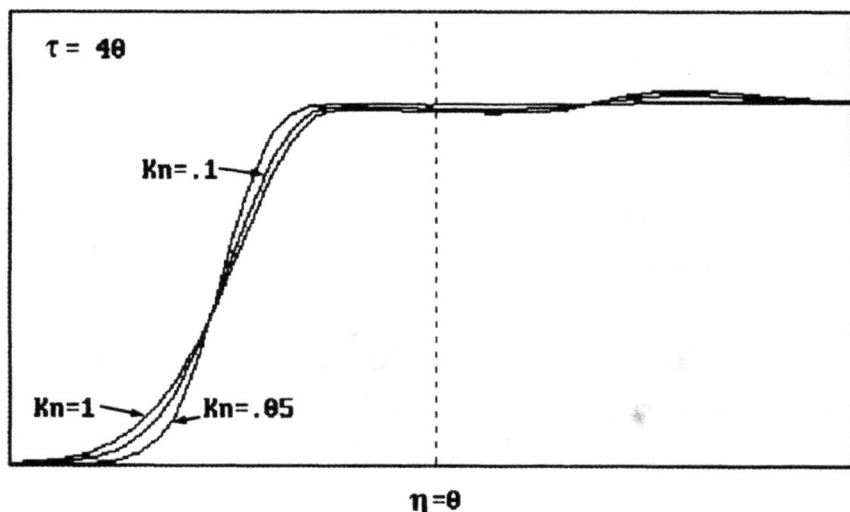

Fig. 4.7 — *Density profiles at different Knudsen numbers*

Numerical density profiles propagating with a constant speed are obtained after a large τ and are shown in Fig. 4.7 for three different values of the Knudsen number Kn and with $Kn_2 = 10^{-5}$ starting from $A_- = b_- = 1$ ($\Longrightarrow \beta = -0.26$).

It is shown that for small Knudsen numbers the shock structure tends to the sharp behaviour of continuum mechanics.

Analogous calculations for gas mixtures and triple collisions are given in [19].

4.5   Discussion and Open Problems

It has been shown, through the development of the content of this chapter, that, despite the relevance of the problem, several interesting and relevant aspects are left open. These open problems can be regarded by the reader as suitable fields for future research activity. The reader must be, however, well aware of the fact that some (if not all) of these problems may be very difficult objects to deal with.

Keeping this in mind, let us first examine the qualitative aspects of the analysis of the problem in one space dimension. As we have seen in Section 4.3, the proof of existence of shock profiles needs first the proof of existence of Euler solutions (strict hyperbolicity and genuine nonlinearity of the Euler equations and uniqueness of the solution to the Rankine-Hugoniot equations), then one can deal with the direct proof of existence of shock profiles.

The proofs of the first step are necessary, but not sufficient, conditions to deal with the proof of the second step. The first step, after the paper by Kawashima and Bellomo [10] can be regarded solved at least for a discrete velocity model with only one velocity modulus. The same technique can hopefully solve the same problem for more general discrete models of the Boltzmann equation.

On the other hand, we have seen that very little is known about Step 4. In fact the few results available in the literature refer to the Broadwell model. Generalizing what is known for the Broadwell model to more general discrete velocity models would certainly be of great interest.

This program could be realized at least for *weak* shock waves. In this case one can use mathematical results which are known for the Navier-Stokes model [20,21]. Then it may be possible to show that for weak shock waves, the profiles described by the discrete Boltzmann equation remain close to the ones predicted by the Navier-Stokes equation. This may require, as already observed in [11], structural assumptions on the discrete velocity models.

Being aware of the fact that the above stated conjectures are still at the stage of speculations without useful ideas towards the solution of the problem, we need mentioning that a similar result is proven for the full Boltzmann equation by Caflisch and Nikolaenko [22].

All speculations presented in this section, refers to problems in one space dimension, as the whole analysis of this chapter. Problems in two or three space dimensions, even if of great interest in fluid-dynamics, are not dealt with, at present, in the pertinent literature. This aspect of the problem certainly deserves future research activity.

References

[1] M. Kogan, **Rarefied Gas Dynamics**, Plenum Press, New York, 1969.

[2] R. Monaco, "Shock wave propagation in gas mixtures by a discrete velocity model of the Boltzmann equation", *Acta Mechanica*, **55**, 1985, p.239.

[3] J. Broadwell, "Shock structure in a simple discrete velocity gas", *Phys. Fluids*, **7**, 1964, p.1243.

[4] R. Gatignol, "Kinetic theory for a discrete velocity gas and application to the shock structure", *Phys. Fluids*, **18**, 1975, p.153.

[5] H. Cabannes, "Etude de la propagation des ondes dans un gaz a quatorze vitesses", *J. de Mecanique*, **14**, 1975, p.705.

[6] R. Caflisch, "Navier-Stokes and Boltzmann shock profiles for a model of gasdynamics", *Comm. Pure Appl. Math.*, **32**, 1979, p.521.

[7] R. Monaco, "On the shock wave structure in binary gas mixtures

for a gas represented by a discrete velocity model", in **Rarefied Gas Dynamics**, Eds. V. Boffi and C. Cercignani, Teubner-Verlag, Stuttgart, Vol.1, 1986, p.245.

[8] R. Monaco, M. Pandolfi Bianchi and T. Platkowski, "Shock-wave formation by the discrete Boltzmann equation for binary gas mixtures", *Acta Mechanica*, **84**, 1990, p.175.

[9] N. Bellomo and E. Longo, "Shock profiles in one dimension by the discrete Boltzmann equation with multiple collisions", in **Waves and Stability in Continuous Media**, *Advances in Mathematics for Applied Science* vol.4, Ed. S. Rionero, World Scientific, London, Singapore, 1991, p.22.

[10] S. Kawashima and N. Bellomo, "On the Euler equations arising in discrete kinetic theory", in **Advances in Kinetic Theory and Continuum Mechanics**, Eds. R. Gatignol and Soubbaramayer , Springer-Verlag, Berlin, New York, 1991, p.73.

[11] S. Kawashima, "Asymptotic behaviour of solutions to the discrete Boltzmann equation", in **Discrete Models of Fluid-Dynamics**, *Advances in Mathematics for Applied Sciences* vol.2, Ed. A. Alves, World Scientific, London, Singapore, p.35.

[12] N. Bellomo and I. Bonzani, "Nonlinear shock waves by the discrete Boltzmann equation", in **Proc. First Int. Conf. Shock Wave Propagation**, Ed. E. Cohen, SIAM, Philadelphia, to appear in 1991.

[13] I. Bonzani and N. Bellomo, "Mathematical aspects of shock wave phenomena by the discrete Boltzmann equation", in **Nonlinear and Dissipative Waves**, Ed. D. Fusco, Pitmann, London, to ap-

pear in 1991.

[14] S. Kawashima and A. Matzumura, "Asymptotic stability of travelling wave solution of systems for one dimensional gas motion", *Comm. Math. Phys.*, **101**, 1985, p.97.

[15] R. Caflisch and Tai Ping Liu, "Stability of shock waves for the Broadwell model", *Comm. Math. Phys.*, **114**, 1988, p.103.

[16] E. Longo and R. Monaco, "On the thermodynamics of the discrete models of the Boltzmann equation for gas mixtures", *Transp. Theory Statist. Phys.*, **17**, 1988, p.423.

[17] B. Hamel, "Disparate mass mixture flows", in **Rarefied Gas Dynamics**, Ed. J. Potter, AIAA, New York, 1977, p.171.

[18] L. Sirovich and E. Goldman, "Normal shock structure in a binary gas", in **Rarefied Gas Dynamics**, Eds. L. Trilling and H. Wachman, Academic Press, New York, 1969, p.407.

[19] M. Pandolfi Bianchi, "Modelling and nonlinear shock waves for binary gas mixtures by the discrete Boltzmann equation with multiple collisions", *Transp. Theory Statist. Phys.*, to appear in 1991.

[20] R. Pego, "Stable viscosities and shock profiles for systems of conservation laws", *Trans. Amer. Math. Soc.*, **282**, 1984, p.749.

[21] A. Maida and R.L. Pego, "Stable viscosity matrices for systems of conservation laws", *J. Diff. Equat.*, **56**, 1985, p.229.

[22] R. Caflisch and B. Nicolaenko, "Shock profile solutions of the Boltzmann equation", *Comm. Math. Phys.*, **86**, 1982, p.161.

CHAPTER 5

INTERNAL AND EXTERNAL FLOWS

As known in gas dynamics [1,2], the analysis of external and internal flows is of relevant interest for the applications. In the first case one deals with the computation of the flow around blunt bodies, airfoils or any spacecraft geometry with the aim of computing the classical aero-thermo-dynamical coefficients: lift, drag and heat transfer coefficients. In the second case one deals with the flow between parallel walls or, more in general, along ducts with varying geometry in order to compute the flow patterns in the duct and the transfer properties on the wall. In both cases the flow may be unsteady (time-space dependent) or steady (time invariant) as this condition may be reached asymptotically in time.

As we shall see, in discrete kinetic theory a crucial point is the mathematical formulation of the boundary conditions at the solid walls. This problem was first dealt with, in a systematic way, by Gatignol [3] and afterwards furtherly developed in [4]. Once such a problem is solved, then one has to tackle an initial-boundary value problem for the kinetic equations in order to compute, in the usual way, the number densities N_i linked to the velocity v_i of the discretized velocity space.

With respect to the full Boltzmann equation, the discrete Boltzmann equation is a relatively simpler object to handle in the applications (as already verified in the preceding chapters) and therefore its study may hopefully lead to useful and easy-to-get practical results.

This idea, however, has still to be regarded as a research program; in fact the applications which are known in the literature still refer to simple problems, namely to computing the flow conditions in the case of simple geometries of the boundary in both internal and external flows. Nevertheless since the aim of this chapter is a perspective one, we shall provide indications, hopefully useful, toward the solution of more sophisticated problems.

Keeping these preliminaries in mind, we can now briefly describe the content of this chapter. The first section deals with the mathematical formulation of the boundary conditions. The second section gives a brief review of the results available in the literature and presents the applications which will be developed in the four remaining sections of this chapter. Section 3 deals with thermal creep problems, Section 4 with Couette flows, Section 5 with Rayleigh flows and finally Section 6 with flows around convex bodies.

5.1   Mathematical Formulation of Boundary Conditions

The pioneer and still fundamental paper on the formulation of the boundary conditions is the one by Gatignol [3], who gave an analysis of the main mathematical and physical constraints for the statement of the boundary conditions. As already mentioned, further analysis is given in [4].

In this section, we shall provide the mathematical formulation of the Maxwell boundary conditions and of the scattering kernel theory.

For sake of simplicity, we shall deal first with a gas between parallel fixed walls at different temperatures in the one-dimensional case. Then the procedure will be generalized to problems with moving walls. Generalizations to problems in more than one space dimension are formally immediate. However it must be mentioned that, as it will be shown in Section 5.2, most of the problems dealt with in the pertinent literature refer to problems in one space dimension.

Keeping this in mind, consider a gas confined, as shown in Fig. 5.1, by two solid parallel walls located at $y = 0$ and $y = \ell$.

The evolution equation is the discrete Boltzmann equation

$$(\frac{\partial}{\partial t} + \mathbf{v}_i \cdot \nabla_{\mathbf{x}})N_i = J_i[\mathbf{N}] \quad , \quad i = 1, \ldots, n \qquad (5.1)$$

which may refer to a simple gas or to a gas mixture. The initial-boundary value problem is an evolution problem such that Eq.(5.1) is joined to the initial data

$$N_{io} = N_i(t = 0, y) \quad , \quad \text{for } y \in [0, \ell] \text{ and } i = 1, \ldots, n$$

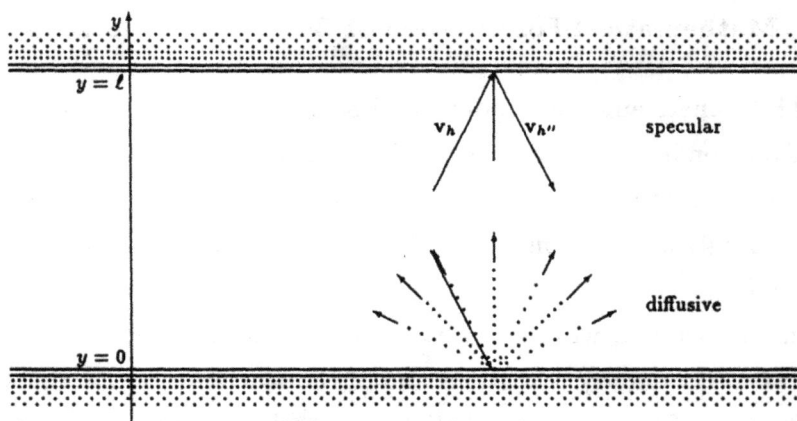

Fig. 5.1 — *Geometry of a gas confined by two parallel walls.*

and to the boundary conditions

$$N_i = N_i(t, y = 0) \ , \quad i = 1, \ldots, n \tag{5.2a}$$

or

$$N_i = N_i(t, y = \ell) \ , \quad i = 1, \ldots, n \tag{5.2b}$$

or, in the most general case, in terms of a map relating the densities leaving the surface to the densities hitting the wall. In this case, considering that one defines on each wall only one half of the velocities, the map must be given on both walls. We will be more precise on this point in what follows.

Moreover, it will be shown that in this case, which is somewhat physically more consistent than the boundary conditions stated in (5.2), the velocity discretization has to obey suitable constraints which are necessary to the formulation of the boundary conditions.

In the framework of classical kinetic theory of fluids [5], the Maxwell boundary conditions are defined as the ones of a gas which is re-emitted by a fixed solid wall either with specular or with diffuse re-emission law.

Considering then the coupling of equations on one of the two solid walls, we assume that such an interaction corresponds to a specular-diffusive reflection which is instantaneous and number preserving.

In order to specify the reflection law the following notation will be needed. Let I be the set of indexes corresponding to the allowed velocities

$$\mathbf{v}_h = \{v_{hx}, v_{hy}, v_{hz}\} \in \mathbf{R}^3 \quad , \quad h \in I = \{1, \ldots, n\} \ .$$

The set I can be partitioned as $I = I^- \cup I^+ \cup I^\circ$ where

$$I^- = \{i \in I : v_{iy} < 0\} \ ,$$

$$I^+ = \{j \in I : v_{jy} > 0\} \ , \qquad (5.3)$$

$$I^\circ = \{h \in I : v_{hy} = 0\} \ .$$

According to this notation we define the vector densities

$$\mathbf{N}^- = \{N_i\}_{i \in I^-} \quad , \quad \mathbf{N}^+ = \{N_j\}_{j \in I^+} \quad , \quad \mathbf{N}^\circ = \{N_h\}_{h \in I^\circ} \ . \quad (5.4)$$

Furthermore we define I_i the set of indexes of incoming particles and I_r the set of indexes of reflected particles.

Thus a particle with velocity with the index in I^- is hitting the wall if it is located at the bottom wall $y = 0$, but is reflecting if $y = \ell$, i.e. $I^- = I_i$ at $y = 0$ and $I^- = I_r$ at $y = \ell$. Trivially the opposite is true for I^+ .

In order to obtain a satisfying reflection law, one needs assuming that the discrete model satisfies the following conditions:

1) $\forall h \in I = \{1, \ldots, n\}$: $v_{hy} \neq 0$

2) $\forall h \in I \quad \exists h' \in I$: $\mathbf{v}_{h'} = -\mathbf{v}_h$

3) $\forall h \in I \quad \exists h'' \in I$: $\mathbf{v}_{h''} = \mathbf{v}_h - 2v_{hy}\mathbf{j}$

4) $\exists h, k \in I$: $|\mathbf{v}_h| \neq |\mathbf{v}_k|$

5) $\boldsymbol{\Psi} = \{\Psi_1, \ldots, \Psi_n\} \in \mathbf{R}^n$ is a collision invariant of the above model if and only if

$$\forall h \in I \qquad \Psi_h = c_\nu + c_x v_{hx} + c_y v_{hy} + c_z v_{hz} + c_T |\mathbf{v}_h|^2 \, , \qquad (5.5)$$

where, as usual, $c_\nu, c_x, c_y, c_z, c_T \in \mathbf{R}$ and do not depend on h but only on t and $\mathbf{x}$. If planar initial–boundary value problems are concerned with, Eq.(5.5) holds with $c_z = 0$.

Condition (1) excludes the possibility of having particles grazing the wall, i.e. $I^\circ = \emptyset$. Condition (3) assures the possibility of specular reflection. Condition (4) allows temperature variation. Finally condition (5) assures that the space of collision invariants has the correct dimension. This allows to characterize a purely diffusive re-emission law in which the reflected gas is Maxwellian and in thermodynamic equilibrium with the wall temperature T_w.

Keeping this in mind, one can characterize, according to [3], the gas-surface interaction on the wall $y = y_w$ (say the bottom one) as

$$\forall j \in I_r \ , \qquad v_{jy} N_j(t, y_w) = \sum_{i \in I_i} -v_{iy} B_{ij}(t) N_i(t, y_w) \qquad (5.6)$$

where $B_{ij} \geq 0$ is the transition probability density that a particle with velocity $\mathbf{v}_i$, $v_{iy} < 0$ hitting the wall is instantaneously reflected with velocity $\mathbf{v}_j$, $v_{jy} > 0$.

Since the reflection law is number preserving, that is the wall does neither absorb nor release particles, B_{ij} must satisfy the normalization condition

$$\forall i \in I_i \ , \qquad \sum_{j \in I_r} B_{ij} = 1 \ . \qquad (5.7)$$

From (5.6) and (5.7) it follows that the total mass flux through the wall is

$$
\begin{aligned}
\nu \mathbf{u}(t, y_w) \cdot \mathbf{j} &= \sum_{h \in I} N_h(t, y_w) v_{hy} = \\
&= \sum_{j \in I_r} N_j(t, y_w) v_{jy} + \sum_{i \in I_i} N_i(t, y_w) v_{iy} = \\
&= \sum_{i \in I_i} \sum_{j \in I_r} -v_{iy} B_{ij} N_i(t, y_w) + \sum_{i \in I_i} N_i(t, y_w) v_{iy} = 0 \ ,
\end{aligned}
\qquad (5.8)
$$

according to the fact that the number of gas particles should be preserved.

If it is assumed that a fraction $\alpha_D(t) \in [0, 1]$ of the molecules is re-emitted diffusively from the surface, while the fraction $1 - \alpha_D(t)$ is reflected in a specular fashion, we can specify the coefficients B_{ij} as

$$B_{ij} = B_{ij}(t, \alpha_D(t)) = [1 - \alpha_D(t)]B_{ij}^S(t) + \alpha_D(t)B_j^D(t) \ , \qquad (5.9)$$

where $B_{ij}^S = B_{ij}(\alpha_D = 0)$ is the probability density of a specular re-emission law and $B_j^D = B_{ij}(\alpha_D = 1)$ is the purely diffusive part. B_j^D does not depend on i because a particle which is reflected diffusively has completely lost the memory of the incoming velocity (see [3]).

If the wall is fixed, recalling condition (3), one can write

$$B_{ij}^S = \delta_{ij''} \qquad \text{with} \qquad \mathbf{v}_j - \mathbf{v}_{j''} = 2v_{jy}\mathbf{j} \qquad (5.10)$$

where $\delta_{ij''}$ is the Kronecker delta.

Equations (5.9) and (5.10) allow then to re-write (5.6) as

$$\forall j \in I_r \qquad v_{jy}N_j = (1 - \alpha_D)N_{j''}v_{jy} + \alpha_D B_j^D \sum_{i \in I_i} -v_{iy}N_i. \qquad (5.11)$$

Since the number of molecules is preserved and the re-emitted gas is in Maxwellian equilibrium with the surface, from (5.6), (5.8) and (5.11) one has the expressions of the diffusive transition rates

$$\forall j \in I_r \qquad B_j^D(t) = \frac{v_{jy}\widehat{N}_j(t, y_w)}{\sum_{k \in I_r} v_{ky}\widehat{N}_k(t, y_w)} \qquad (5.12)$$

where $\widehat{N}_k$, $k \in I_r$ are the densities in an absolute Maxwellian equilibrium with temperature T_w.

Referring to Section 1.4.1, a standard algebraic manipulation shows that the absolute Maxwellian equilibrium densities have $c_x = c_y = c_z = 0$, and hence from Eq.(5.5)

$$\widehat{N}_h(t,y) = A(t,y)\exp[c_T(t,y)|\mathbf{v}_h|^2] \ , \quad h \in I \ , \tag{5.13}$$

where $A = e^{c_\nu}$.

Therefore in this case and for models with more than one velocity modulus, temperature is given by

$$\widehat{T}(y,t) = \frac{1}{3\mathcal{R}} \frac{\displaystyle\sum_{h\in I} \mathbf{v}_h^2 \exp(c_T|\mathbf{v}_h|^2)}{\displaystyle\sum_{h\in I} \exp(c_T|\mathbf{v}_h|^2)} \tag{5.14}$$

where $\mathcal{R} = k_B/m$ is the gas constant.

As already shown in Section 1.4.1, the jacobian of the transformation relating the Maxwellian parameters to the independent macroscopic observables never vanishes and, therefore, Eq.(5.14) defines a one-to-one mapping that can be formally inverted as

$$c_T = g(\widehat{T}) \ . \tag{5.15}$$

Thus by Eqs.(5.13) and (5.15), Eq.(5.12) may be re-written as

$$B_j^D(T_w) = \frac{v_{jy}\exp[g(T_w)|\mathbf{v}_j|^2]}{\displaystyle\sum_{k\in I_r} v_{ky}\exp[g(T_w)|\mathbf{v}_k|^2]} \tag{5.16}$$

with $T_w = T_w(t)$ given by the boundary conditions.

This explicitly gives the dependence of B_j^D from the wall temperature that should be substituted in the reflection law (5.11). Thus (5.11) defines in y_w the density N_j , $j \in I_r$ of the reflecting particles in terms of the densities N_i , $i \in I_i$ of the impinging ones, of the wall temperature T_w and of the re-emission parameter $\alpha_D(t)$ which depends on the structural and thermodynamic characteristics of the wall. When α_D is given, the mathematical formulation is possible.

Paper [4] has shown how one can exploit the continuity of the heat flux at the wall (when this is known) to compute α_D, which otherwise has to be assumed a priori.

Consider now the problem with *moving* boundaries and, in particular, the case in which the walls may move with a velocity $w_w = w_w(t)$ along the x–axis, so that their distance remains unchanged.

We note that in this case specular reflection boundary conditions cannot be used, since, because of the interaction with the wall, particles would assume a speed not allowed by the selected discretization of the velocity space. Thus, in what follows, α_D in (5.9) has to be set equal to one. In this case the geometrical conditions to be satisfied are (1), (2) and (5) where the absolute Maxwellian equilibrium densities are given by

$$\widehat{N}_h(t,y) = A(t,y) \exp[c_x(t,y)v_{hx} + c_y(t,y)v_{hy} + c_T(t,y)|\mathbf{v}_h|^2] . \quad (5.17)$$

For sake of completeness, consider the more general case of models with more than one velocity modulus. Formulae for models with a single velocity modulus are easily obtained by setting $c_T = 0$.

The equilibrium drift velocity and temperature are given by

$$\widehat{\mathbf{u}}(y,t) = \frac{\displaystyle\sum_{h\in I} \mathbf{v}_h \exp(c_x v_{hx} + c_y v_{hy} + c_T |\mathbf{v}_h|^2)}{\displaystyle\sum_{h\in I} \exp(c_x v_{hx} + c_y v_{hy} + c_T |\mathbf{v}_h|^2)} \tag{5.18a}$$

$$\widehat{T}(y,t) = \frac{1}{3\mathcal{R}} \left(\frac{\displaystyle\sum_{h\in I} |\mathbf{v}_h|^2 \exp(c_x v_{hx} + c_y v_{hy} + c_T |\mathbf{v}_h|^2)}{\displaystyle\sum_{h\in I} \exp(c_x v_{hx} + c_y v_{hy} + c_T |\mathbf{v}_h|^2)} - u_x^2 - u_y^2 \right) . \tag{5.18b}$$

As already shown in Section 1.4.1, the jacobian of the transformation in (5.18), as the one in (5.14), never vanishes. Therefore Eqs.(5.18) define a one-to-one mapping $(c_x, c_y, c_T) \mapsto (\widehat{u}_x, \widehat{u}_y, \widehat{T})$ that can be formally inverted as

$$c_x = g_1(\widehat{u}_x, \widehat{u}_y, \widehat{T})$$

$$c_y = g_2(\widehat{u}_x, \widehat{u}_y, \widehat{T}) \tag{5.19}$$

$$c_T = g_3(\widehat{u}_x, \widehat{u}_y, \widehat{T}) .$$

Thus by (5.8), (5.17) and (5.19), one can re-write Eq.(5.12) as

$$B_j^D(w_w, T_w) = \frac{v_{jy} \exp[G_j(w_w, 0, T_w)]}{\displaystyle\sum_{k\in I_r} v_{ky} \exp[G_k(w_w, 0, T_w]} \tag{5.20}$$

where

$$G_j(w_w, 0, T_w) = g_1(w_w, 0, T_w)v_{jx} + g_2(w_w, 0, T_w)v_{jy} + g_3(w_w, 0, T_w)|\mathbf{v}_j|^2$$

with $T_w = T_w(t)$ and velocity $w_w = w_w(t)$ of the moving wall along the x–axis defined by the boundary conditions themselves.

Similarly in isothermic problems dealt with by single modulus models, the relation between B_j^D and w_w given by (5.18) may be inverted and therefore (5.20) still holds with $g_3 \equiv 0$ and the elimination of the temperature dependence.

Consider now that the boundary $\partial\Omega$ is not a plane and let $\mathbf{n}$ be the inward-pointing (i.e. into the gas) unit vector normal to the surface in $P \in \partial\Omega$ and $\mathbf{w}_w$ be the velocity of the wall in P. In this case equation (5.6) re-writes

$$\forall j \in I_r \qquad (\mathbf{v}_j - \mathbf{w}_w) \cdot \mathbf{n} N_j = \sum_{i \in I_i} -(\mathbf{v}_i - \mathbf{w}_w) \cdot \mathbf{n} B_{ij} N_i \qquad (5.21)$$

where now I_i is the set of indexes such that $(\mathbf{v}_i - \mathbf{w}_w) \cdot \mathbf{n} < 0$, i.e. the indexes related to the particles which can collide with the surface. On the other hand I_r is the set of indexes such that $(\mathbf{v}_j - \mathbf{w}_w) \cdot \mathbf{n} > 0$. From (5.21) it again follows that

$$\nu(\mathbf{u} - \mathbf{w}_w) \cdot \mathbf{n} = \sum_{j \in I_r} N_j (\mathbf{v}_j - \mathbf{w}_w) \cdot \mathbf{n} + \sum_{i \in I_i} N_i (\mathbf{v}_i - \mathbf{w}_w) \cdot \mathbf{n}$$

$$= \sum_{j \in I_r} \sum_{i \in I_i} -N_i B_{ij} (\mathbf{v}_i - \mathbf{w}_w) \cdot \mathbf{n} + \sum_{i \in I_i} N_i (\mathbf{v}_i - \mathbf{w}_w) \cdot \mathbf{n}$$

$$= -\sum_{i \in I_i} N_i (\mathbf{v}_i - \mathbf{w}_w) \cdot \mathbf{n} \sum_{j \in I_r} B_{ij} + \sum_{i \in I_i} N_i (\mathbf{v}_i - \mathbf{w}_w) \cdot \mathbf{n} \,,$$

which vanishes according to (5.7).

As far as the diffusive part is concerned, Eq.(5.12) re-writes

$$\forall j \in I_r \qquad B_j^D = \frac{(\mathbf{v}_j - \mathbf{w}_w) \cdot \mathbf{n} N_j}{\sum_{k \in I_r} (\mathbf{v}_k - \mathbf{w}_w) \cdot \mathbf{n} N_k} .$$

It is possible to provide a formulation of the boundary conditions more general than the one just given exploiting the expression (5.6) without the limitation of the form of the coefficients B_{ij} to the specular or diffusive case. This formulation corresponds to the analogous one (the so-called scattering kernel theory) widely used for the full Boltzmann equation, see ref.[6].

The main difficulty in using this type of formulation consists in the fact that no mathematical model is available in the literature for expressions of B_{ij} different than the ones corresponding to specular reflection or diffuse re-emission or a combination of the two.

Even for the full Boltzmann equation comparisons between theory and experiment have shown a difficulty, or even impossibility, in matching experimental data with the prediction of theoretical results available in the literature [7,8].

As a matter of fact all applications in fluid-dynamics deal, as we shall see, with a Maxwell type formulation of the boundary conditions.

5.2   Applications of the Discrete Boltzmann Equation in Bounded Domains

The aim of this chapter is to transfer at a practical level the theoretical analysis just developed in the preceding section, applying it to study several flow configurations.

Before going into the details of the above mentioned topics, a survey of the various results (both mathematical and applicative) available in the literature can be provided. Essentially, the research activities are directed toward three main directions:

i) Proving mathematical results on existence, uniqueness and stability of solutions to initial-boundary value problems.

ii) Studying analytical solutions (generally for simple models applied to the analysis of special initial-boundary value problems).

iii) Giving numerical solutions for general flow conditions studied by the application of quite general models.

The mathematical aspects of the initial-boundary value problem have been mainly studied by Kawashima [9,10], who was able to prove global existence and uniqueness of the solutions for a gas between parallel walls reflecting diffusively with large initial conditions for a model with binary collisions. Kawashima's result is obtained extending, by simple geometric arguments, the results on the initial value problem on the whole real axis $\mathbb{R}$ (already reviewed in Chapter 3) to the analysis of the initial-boundary value problem. Kawashima's paper also deals with stability analysis. Additional results are obtained in [11,12].

Moreover, mathematical aspects of the steady solutions for the plane Broadwell model for a simple gas in two dimensions are dealt with in [13].

Analytical solutions to various problems in fluid-dynamics (internal Couette flows and Rayleigh flows) have been obtained by several authors. If the mathematical model of the discrete Boltzmann equation is a very simple one, say the 4–velocity planar model, then it is rather simple to obtain analytical solutions as shown in the pioneer work by Broadwell [14]. Broadwell's analysis has been extended to more general physical conditions by Cabannes [15] on the basis of his 14–velocity model. Analytical solutions to the steady flow of a binary gas mixture modelled by the planar 2×4–velocity model has been obtained by Longo [16].

Still dealing with a survey of analytical solutions we need mentioning the solution of the Rayleigh flow for a binary gas mixture obtained in [17] using again the planar 2×4–velocity model. Additional analytical solutions are due to Cornille, see ref.[18] and the wide bibliography quoted therein, which are based upon the search for solitons and bisolitons to obtain analytical solutions to the initial-boundary value problem, in a fashion analogous to the one already reviewed in Chapter 3.

The solution of the unsteady Couette flow has been obtained by numerical techniques by Gatignol [19] for the planar 4–velocity model for a simple gas. This result has been extended in [20] where both binary and triple collisions for the planar 6–velocity model are considered. This last paper shows two important aspects of the problem referred to the 6–velocity model:

— The relevance of dealing with triple collisions is referred to the need of defining uniquely the Maxwellian state. As a matter of fact such a model, if triple collisions are added to binary collisions, satisfies conditions (1)–(3) and (5) defined in Section 5.1 (in order to satisfy also conditions (4) a model with two velocity moduli should be necessary).

— The influence of the triple collision term on the qualitative and quantitative behaviour of the flow. As a matter of fact this term, which can hopefully simulate some dense gas effects, shows how the trend of the gas towards the steady state flow is obtained faster with the decreasing of the Knudsen number, namely with increasing influence of the triple collision term.

The same line of thought is shared in [21] where thermal creep problems between moving walls are studied by the planar 2×6–velocity model with triple collisions.

In the sections which follow, several flow configurations will be studied. Of course, not all physical situations will be covered, but however the analysis of some classical flows will be dealt with as a trial ground, leaving to the reader further generalization or developments. In all cases one has to keep in mind that some of these generalizations are not immediate and need the support of a detailed research activity.

In details we shall deal with the mathematical formulation of the following problems:

— Flow between two parallel walls at different temperatures;

— Couette flow;

— Rayleigh flow;

— Flow of a Maxwellian stream over two-dimensional bodies.

Three applications will also be presented with reference to the first three problems. No meaningful application is known, as we shall see, with regard to the fourth problem, which can be seen by the reader as a conceivable field for future research activity.

5.3 Flow between Two Walls at Different Temperatures

Consider now a gas confined by two parallel fixed plates at different temperatures. The mathematical formulation of such a problem is the one given in Section 5.1. Namely one has an initial-boundary value problem with boundary conditions stated as in (5.6), where the terms B_{ij} may also be determined in terms of partially diffusive, partially specular reflection or, more in general, in the framework of the scattering kernel theory.

Classically, solving the initial-boundary value problem provides the densities $N(t, y)$ between the two walls. One can then recover the macroscopic observables. In particular, from (5.8) the mass flow in the direction of the y–axis must vanish at the walls. On the other hand, as confirmed by experimental evidence, the temperature profiles will not reach exactly the temperature of the wall. In fact the temperature of the gas near the wall is a weighted average of the temperature of the particles that are going to strike the wall, of those reflected specularly and of those reflected diffusively. Only those belonging to this last category have the temperature of the wall.

Then it makes sense to consider the differences between the gas temperature at $y = 0$ and $y = \ell$, say $T(t, 0)$ and $T(t, \ell)$, and the corresponding temperatures of the walls, $T_{wo}(t)$ and $T_{w1}(t)$ respectively, which define the so-called temperature jump

$$\Delta_{T_o}(t) = T(t, 0) - T_{wo}(t) \quad , \quad \Delta_{T_1}(t) = T(t, \ell) - T_{w1}(t) .$$

In the literature of the continuous Boltzmann equation (see for instance the book [1], Chap.5), it is given an argument suggesting that Δ_T is proportional to $\frac{\partial T}{\partial y}$.

In order to deal with a specific application, we have to select a mathematical model of the discrete Boltzmann equation which satisfies the conditions (1)–(5).

In particular, according to the analysis developed in Section 5.1, we need a model such that

i) The model is characterized by two velocity moduli in order to describe different temperature situations;

ii) The model has sufficient velocity directions to possess specular reflections;

iii) The model can be suitably oriented to avoid grazing directions;

iv) The number of velocities and directions is such that the space of collision invariants has the correct dimension in order to have a unique description of the Maxwellian state.

A mathematical model which satisfies the above conditions is the discrete Boltzmann equation with 6 velocity directions in the plane

$$\mathbf{e}_k = \cos\left[(2k+1)\frac{\pi}{6}\right]\mathbf{i} + \sin\left[(2k+1)\frac{\pi}{6}\right]\mathbf{j} \ , \quad k = 1,\ldots,6 \qquad (5.22)$$

and two velocity moduli c and $2c$, which has been described in Section 2.1.5. The only difference between the model described in Section 2.1.5 and the one used here is that the x and y–axis are oriented differently in order to satisfy (ii)–(iii), i.e. conditions (1) and (3) of Section 5.1. The physical and geometrical settings are then sketched in Fig. 5.2. The collisional scheme and the thermodynamics of the model are, of course, the ones reported in Section 2.1.5.

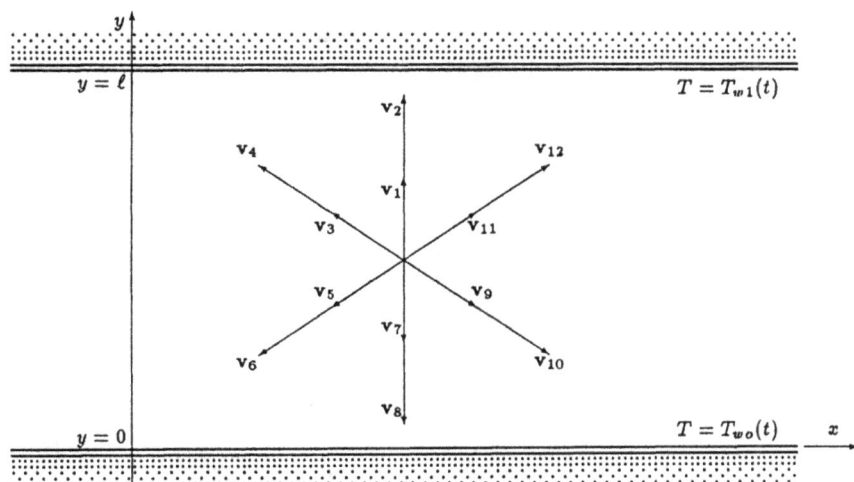

Fig. 5.2 — *Sketch of the physical problem and of the discrete model*

Keeping in mind the change in the coordinate axis, the relation

(2.21) between the Maxwellian parameters and the macroscopic observables can be written as

$$\frac{\widehat{\nu}}{c^2}\left(\frac{4c^2 - \widehat{u}_x^2 - \widehat{u}_y^2}{3} - \mathcal{R}\widehat{T}\right) = 2A(\cosh c_y + 2\cosh\frac{\sqrt{3}}{2}c_x\cosh\frac{c_y}{2})$$

$$\frac{\widehat{\nu}}{c^2}\left(\mathcal{R}\widehat{T} - \frac{c^2 - \widehat{u}_x^2 - \widehat{u}_y^2}{3}\right) = 2B(\cosh 2c_y + 2\cosh\sqrt{3}c_x\cosh c_y)$$

$$\frac{\widehat{\nu}\,\widehat{u}_x}{c} = 2\sqrt{3}A\sinh\frac{\sqrt{3}}{2}c_x\cosh\frac{c_y}{2} + 4\sqrt{3}B\sinh\sqrt{3}c_x\cosh c_y$$

$$\frac{\widehat{\nu}\,\widehat{u}_y}{c} = 2A(\sinh c_y + \cosh\frac{\sqrt{3}}{2}c_x\sinh\frac{c_y}{2})$$

$$+ 4B(\sinh 2c_y + \cosh\sqrt{3}c_x\sinh c_y) .$$

$$(5.23)$$

The fact that the reflected gas is in Maxwellian equilibrium with the surface means that, from (5.8),

$$\widehat{u}_y = 0 \iff c_y = 0$$

and hence, from (5.23), c_x is implicitly given by

$$\sqrt{3}cw_w = \frac{(4c^2 - w_w^2 - 3\mathcal{R}T_w)\sinh\frac{\sqrt{3}}{2}c_x}{1 + 2\cosh\frac{\sqrt{3}}{2}c_x}$$

$$+ 2\frac{(3\mathcal{R}T_w - c^2 + w_w^2)\sinh\sqrt{3}c_x}{1 + 2\cosh\sqrt{3}c_x} ,$$

$$(5.24)$$

where T_w and w_w are respectively the wall temperature and velocity.

The other two Maxwellian parameters are then explicitly given by

$$A = \frac{\hat{\nu}}{6c^2} \frac{4c^2 - w_w^2 - 3\mathcal{R}T_w}{1 + 2\cosh\frac{\sqrt{3}}{2}c_x}$$

$$B = \frac{\hat{\nu}}{6c^2} \frac{3\mathcal{R}T_w - c^2 + w_w^2}{1 + 2\cosh\sqrt{3}c_x} .$$

If the walls are fixed $w_w = 0$, then the solution of Eq.(5.24) is $c_x = 0$ and therefore the absolute Maxwellian number densities are

$$\widehat{N}_{2k-1} = \frac{\hat{\nu}}{6}(\frac{4}{3} - \frac{\mathcal{R}\widehat{T}}{c^2}) \quad , \quad \widehat{N}_{2k} = \frac{\hat{\nu}}{6}(\frac{\mathcal{R}\widehat{T}}{c^2} - \frac{1}{3}) \quad , \quad k = 1, \ldots, 6 . \quad (5.25)$$

Consider now the one-dimensional case of the above model defined by

$$N_3 = N_{11} \quad , \quad N_4 = N_{12} \quad , \quad N_5 = N_9 \quad , \quad N_6 = N_{10}$$

where $N_h = N_h(t, y)$, $h = 1, \ldots, 12$.

The equations can be written in non-dimensional form by the introduction of the quantities

$$[\eta, \tau, \mathbf{N}', T', \mathbf{q}'] = [\frac{y}{\ell}, \frac{ct}{\ell}, \frac{\mathbf{N}}{\nu_o}, \frac{T}{T_o}, \frac{\mathbf{q}}{mc^3\nu_o}] \qquad (5.26)$$

where T_o and ν_o are, respectively, the initial temperature and density of the gas, say in $y = 0$, ℓ is the distance between the walls and $\mathbf{q}$ is the heat flux.

The Maxwellian densities can then be written in this dimensionless form as

$$\widehat{N}'_{2k-1} = \frac{1}{18}(4 - \omega\widehat{T}') \quad , \quad \widehat{N}'_{2k} = \frac{1}{18}(\omega\widehat{T}' - 1) \quad , \quad k = 1, \ldots, 6$$

$$(5.27)$$

where $\omega = \dfrac{3\mathcal{R}T_o}{c^2}$.

It is plain from (5.27) that $\widehat{T}'$ can only range between $\frac{1}{\omega}$ and $\frac{4}{\omega}$, where the lower (respectively higher) temperature is achieved when all particles have the smallest (respectively largest) allowed velocity modulus. Hence the parameter ω (or equivalently the modelling velocity c) has to be chosen so that the discrete velocity model can describe the full range of temperatures required by the physical problem at hand.

The evolution equations for the densities are

$$(\frac{\partial}{\partial\tau} + \frac{\partial}{\partial\eta})N'_1 = 2A_1[\mathbf{N}'] + 2E_1[\mathbf{N}'] + A_2[\mathbf{N}']$$

$$(\frac{\partial}{\partial\tau} + 2\frac{\partial}{\partial\eta})N'_2 = 2B_1[\mathbf{N}'] - 2D_1[\mathbf{N}'] + B_2[\mathbf{N}']$$

$$(\frac{\partial}{\partial\tau} + \frac{1}{2}\frac{\partial}{\partial\eta})N'_3 = -A_1[\mathbf{N}'] + C_1[\mathbf{N}'] + D_1[\mathbf{N}'] - A_2[\mathbf{N}']$$

$$(\frac{\partial}{\partial\tau} + \frac{\partial}{\partial\eta})N'_4 = -B_1[\mathbf{N}'] - E_1[\mathbf{N}'] + D_1[\mathbf{N}'] - B_2[\mathbf{N}']$$

$$(5.28)$$

$$(\frac{\partial}{\partial\tau} - \frac{1}{2}\frac{\partial}{\partial\eta})N'_5 = -A_1[\mathbf{N}'] - C_1[\mathbf{N}'] - D_1[\mathbf{N}'] + A_2[\mathbf{N}']$$

$$(\frac{\partial}{\partial\tau} - \frac{\partial}{\partial\eta})N'_6 = -B_1[\mathbf{N}'] + E_1[\mathbf{N}'] - C_1[\mathbf{N}'] + B_2[\mathbf{N}']$$

$$(\frac{\partial}{\partial\tau} - \frac{\partial}{\partial\eta})N'_7 = 2A_1[\mathbf{N}'] - 2E_1[\mathbf{N}'] - A_2[\mathbf{N}']$$

$$(\frac{\partial}{\partial\tau} - 2\frac{\partial}{\partial\eta})N'_8 = 2B_1[\mathbf{N}'] + 2C_1[\mathbf{N}'] - B_2[\mathbf{N}']$$

where

$$A_1[\mathbf{N}'] = \frac{2}{3Kn_1}(N_3'N_5' - N_1'N_7') \quad , \quad B_1[\mathbf{N}'] = \frac{4}{3Kn_1}(N_4'N_6' - N_2'N_8') \quad ,$$

$$A_2[\mathbf{N}'] = \frac{1}{Kn_2}(N_3'^2 N_7' - N_1'N_5'^2) \quad , \quad B_2[\mathbf{N}'] = \frac{2}{Kn_2}(N_4'^2 N_8' - N_2'N_6'^2) \quad ,$$

$$C_1[\mathbf{N}'] = \frac{\sqrt{7}}{2Kn_1}(N_5'N_6' - N_3'N_8') \quad , \quad D_1[\mathbf{N}'] = \frac{\sqrt{7}}{2Kn_1}(N_5'N_2' - N_3'N_4') \quad ,$$

$$E_1[\mathbf{N}'] = \frac{\sqrt{7}}{2Kn_1}(N_7'N_4' - N_1'N_6') \quad ,$$

$$(5.29)$$

and

$$Kn_1 = \frac{1}{\ell S \nu_o} \quad , \quad Kn_2 = \frac{4\sqrt{\pi}}{5}\frac{\ell}{\sqrt{S}}Kn_1^2$$

which can be considered proportional to two Knudsen numbers, one typical of binary collisions, the other referred to triple collisions.

These equations must be coupled with the boundary conditions (5.11) where, according to (5.12) and (5.27), the diffuse re-emission rates $B_j^D = B_j^D(T_{wo}')$ at the wall $\eta = 0$ are given by

$$B_1^D = \frac{4 - \omega T_{wo}'}{2(\omega T_{wo}' + 2)}$$

$$B_2^D = \frac{\omega T_{wo}' - 1}{\omega T_{wo}' + 2}$$

$$B_3^D = B_{11}^D = \frac{4 - \omega T_{wo}'}{4(\omega T_{wo}' + 2)} \qquad (5.30)$$

$$B_4^D = B_{12}^D = \frac{\omega T_{wo}' - 1}{2(\omega T_{wo}' + 2)} \quad .$$

Therefore

$$N_1' = (1 - \alpha)N_7' + \alpha \frac{4 - \omega T_{wo}'}{2(\omega T_{wo}' + 2)}(N_5' + 2N_6' + N_7' + 2N_8')$$

$$N_2' = (1 - \alpha)N_8' + \alpha \frac{\omega T_{wo}' - 1}{2(\omega T_{wo}' + 2)}(N_5' + 2N_6' + N_7' + 2N_8')$$

$$N_3' = (1 - \alpha)N_5' + \alpha \frac{4 - \omega T_{wo}'}{2(\omega T_{wo}' + 2)}(N_5' + 2N_6' + N_7' + 2N_8') \tag{5.31}$$

$$N_4' = (1 - \alpha)N_6' + \alpha \frac{\omega T_{wo}' - 1}{2(\omega T_{wo}' + 2)}(N_5' + 2N_6' + N_7' + 2N_8') \ .$$

It can immediately be realized that the re-emission coefficients $B_j^D(T_{w1}')$ and the reflection law in $\eta = 1$ are obtained substituting in Eqs.(5.30) and (5.31) the *specular* transformation on the subscripts: $1 \leftrightarrow 7$, $2 \leftrightarrow 8$, $3 \leftrightarrow 5$, $4 \leftrightarrow 6$ and of course T_{w1}' to T_{wo}' .

Consider a problem in which the gas is initially in a constant Maxwellian state with temperature $T'(\tau = 0, \eta) = 1$ and assume that the temperature is achieved by the following distribution of number densities

$$\widehat{N}_i'(\tau = 0) = \frac{1}{12} \ , \quad i = 1, \ldots, 8 \ , \tag{5.32}$$

which corresponds to choosing in (5.27) $\omega = \frac{5}{2}$, or, equivalently, $c = \sqrt{\frac{6}{5}\mathcal{R}T_o}$. Finally let the walls be at temperature

$$T_{wo}'(\tau) = 1 \quad T_{w1}'(\tau) = 1 + 0.1H(\tau) \ , \tag{5.33}$$

where $H(\tau)$ is the Heaviside function, and let them reflect diffusively, i.e. $\alpha_D = 1$ in the boundary conditions (5.31).

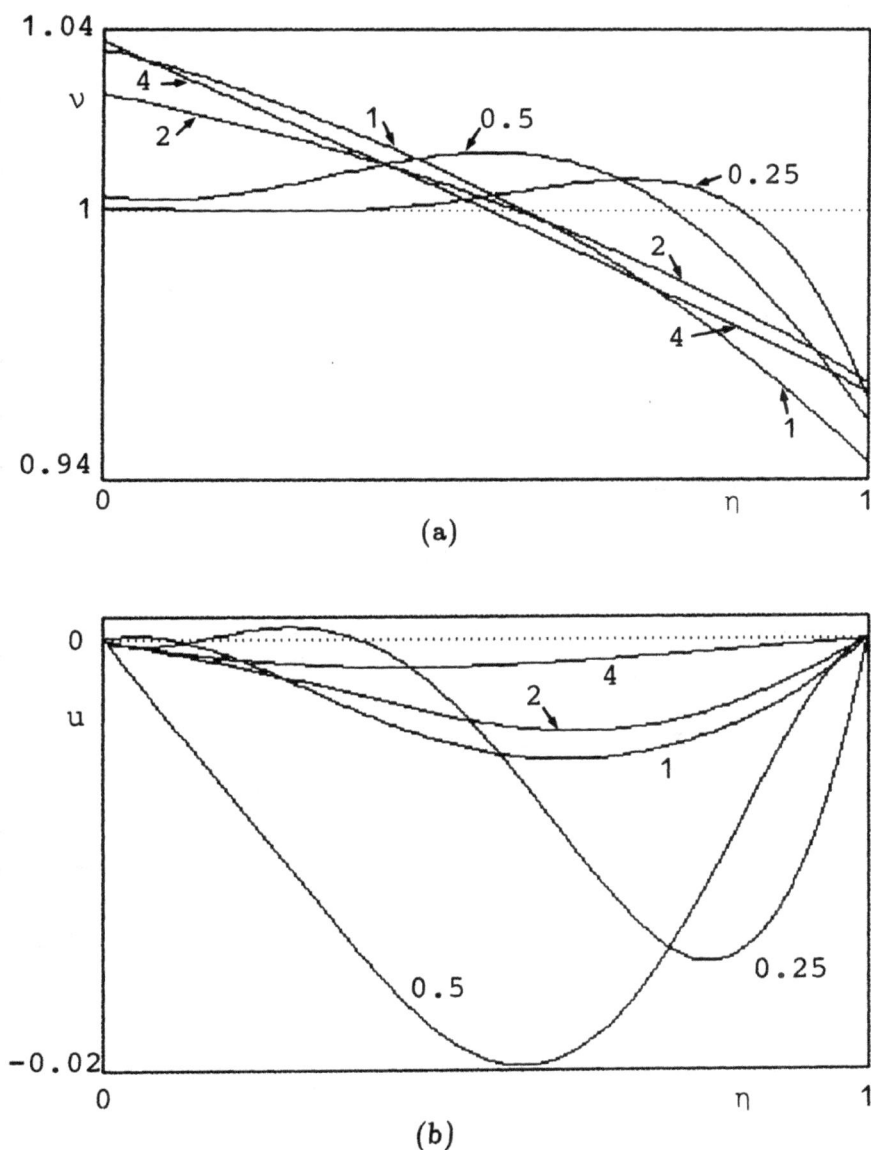

Fig. 5.3 — Behaviour of the macroscopic variables for $Kn_1 = 0.1$ and $Kn_2 = 10^4$ at different times: (a) density; (b) drift velocity

(a)

(b)

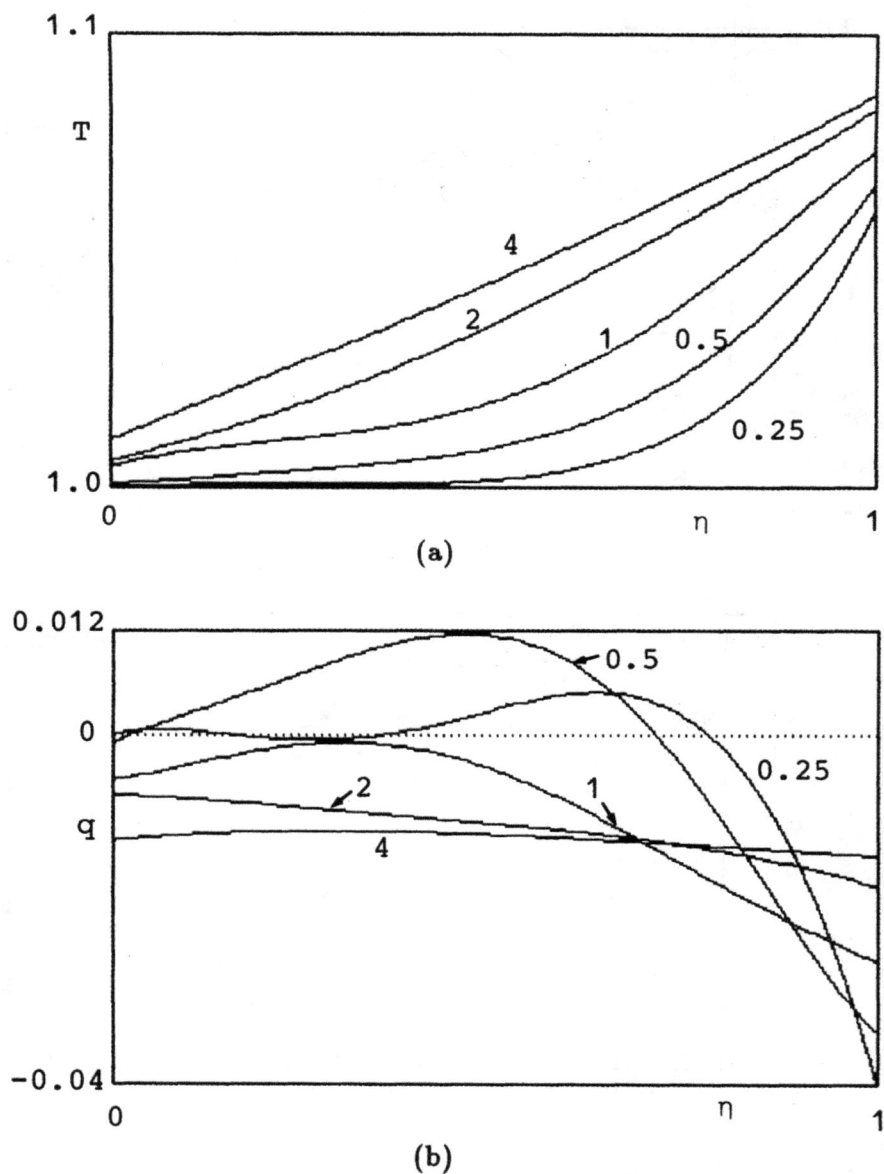

Fig. 5.4 — *Behaviour of the macroscopic variables for $Kn_1 = 0.1$ and $Kn_2 = 10^4$ at different times: (a) temperature; (b) heat flux*

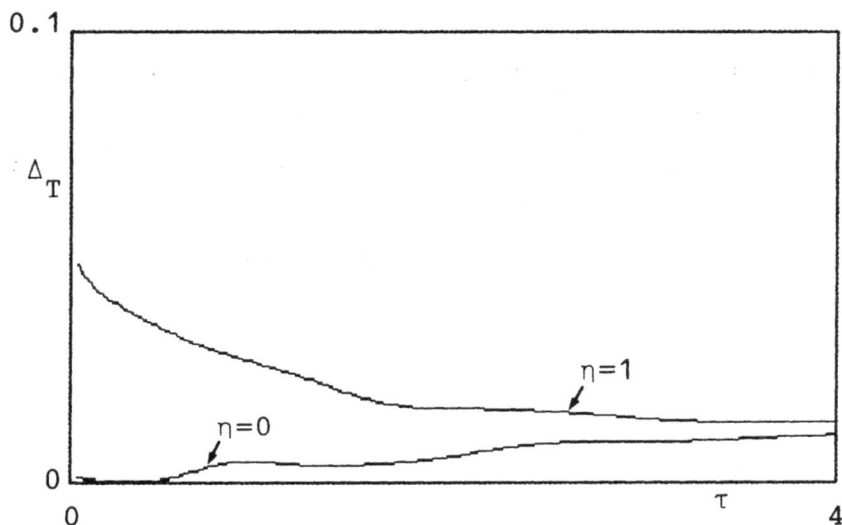

Fig. 5.5 — *Absolute values of the temperature jump at $\eta = 0$ (Δ_{T_o}), $\eta = 1$ (Δ_{T_1}) versus time for $Kn_1 = 0.1$ and $Kn_2 = 10^4$*

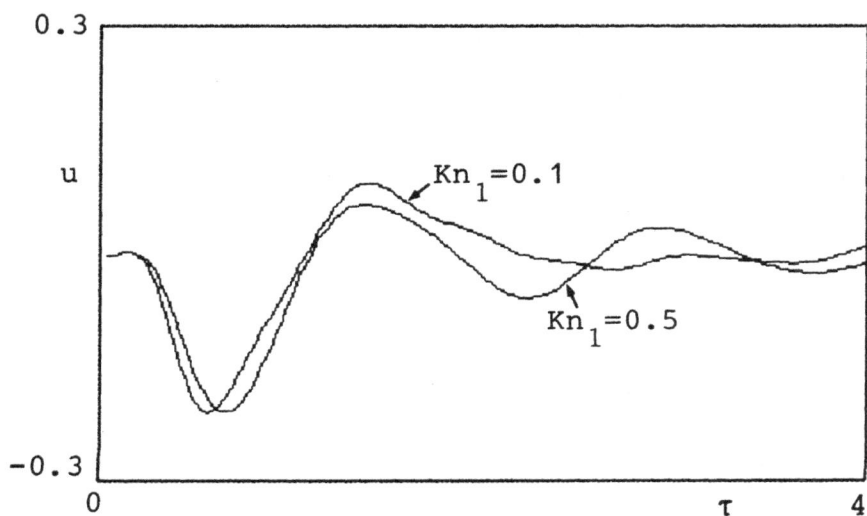

Fig. 5.6 — *Drift in $\eta = 0.4$ versus time for different Knudsen numbers*

The evolution of the macroscopic variables is plotted in Figs. 5.3–5.4. According to an initial negative drift velocity particles crowd near the colder wall. At smaller times, say $\tau = 0.25$, only the particles closer to the warmer wall fill the change in the thermodynamic state while particle near the colder wall hardly move. Hence the minimum of the velocity drift is reached near the hot wall. Then, say at $\tau = 0.5$, the whole gas moves to reach a new equilibrium state. At longer times the drift velocity decreases showing a trend to a steady state characterized by a linear dependence of both density and temperature on η while the heat flux tends to a negative constant.

The evolution of the temperature jump at the walls is already shown in Fig. 5.4a. Its behaviour as a function of time is represented in Fig. 5.5. It should be observed how the absolute values of Δ_T in $\eta = 0$ (Δ_{T_o}) and in $\eta = 1$ (Δ_{T_1}) tend to almost the same asymptotic value.

In Fig. 5.6 the drift velocity is plotted versus time at $\eta = 0.4$ for two different values of Kn_1 $(Kn_2 = 10^6 Kn_1^2)$. A lower Kn_1 corresponds to a denser (and therefore more viscous) gas for which the triple collision term may become important. It should be observed that the lower the Knudsen number is, the faster the steady state is reached.

These figures have been computed numerically using the differential quadrature method fully explained in the second section of Chapter 3.

5.4   Couette Flow

Couette flow is a classical problem in fluid-dynamics, which has also been studied in continuous kinetic theory, see ref.[1], Chap.4. In this case one has to deal with a rarefied gas confined between two parallel walls, as shown in Fig. 5.7. The two walls may not have the same temperature. An unsteady problem is such that at $t = 0$ the gas is in absolute equilibrium with given densities

$$\widehat{N}_{io} = \widehat{N}_i(t = 0, y) = \text{const.} \tag{5.34}$$

between two plates at rest. The lower wall suddenly moves for $t > 0$ with velocity

$$w_w = w_w(t) \tag{5.35}$$

in the direction of the x-axis, as indicated in Fig. 5.7.

If $w_w(t)$ tends to a constant as $t \to \infty$, then the solution of the initial-boundary value problem should lead, for $t \to \infty$, to a steady flow with y–profiles of the macroscopic quantities constant with respect to time as it will be shown in the application.

The unsteady Couette flow problem was studied for the first time in the framework of the discrete Boltzmann equation by Gatignol [19] using the planar 4–velocity model of Section 2.1.1. Here, some details of the solution of such a problem will be given for the regular planar model with 6–velocities and triple collisions.

Consider, as an application, the unsteady problem with both walls at the same temperature and with the boundary conditions (5.35). According to Section 5.1 a mathematical model of the discrete Boltzmann equation fulfilling conditions (1), (2) and (5) is needed. In this case a velocity discretization characterized by a single velocity modulus is sufficient to describe the behaviour of the system.

Therefore the plane regular 6–velocity model with both binary and symmetric triple collisions is selected (see Section 2.1.4). Moreover, in order to be consistent with assumption (1), the model is oriented as indicated in Fig. 5.7.

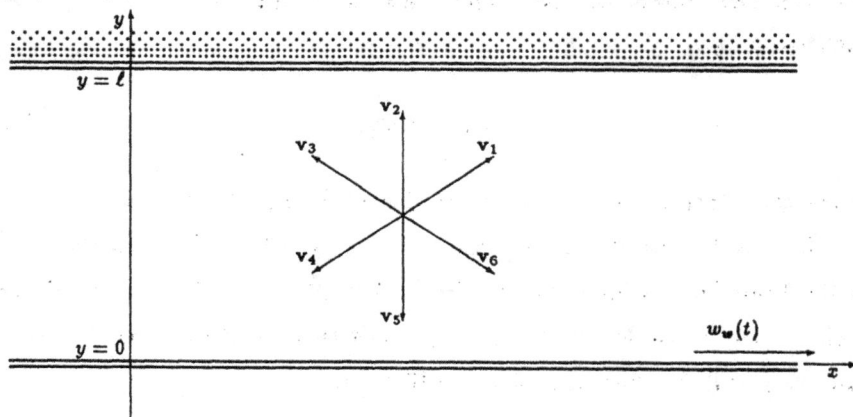

Fig. 5.7 — *Sketch of the physical problem and of the discrete model*

The six velocities corresponding to this discretization are

$$\mathbf{v}_i = c\left[\cos(2i-1)\frac{\pi}{6}\mathbf{i} + \sin(2i-1)\frac{\pi}{6}\mathbf{j}\right] \quad , \quad i = 1,\ldots,6 . \qquad (5.36)$$

Besides the binary head-on collisions

$$(\mathbf{v}_1,\mathbf{v}_4) \longleftrightarrow (\mathbf{v}_2,\mathbf{v}_5) \longleftrightarrow (\mathbf{v}_3,\mathbf{v}_6) \quad ,$$

in order to satisfy condition (5) (with $c_T = 0$ of course), namely to have the right dimension for the space of collision invariants, the triple collisions are included

$$(\mathbf{v}_1, \mathbf{v}_3, \mathbf{v}_5) \longleftrightarrow (\mathbf{v}_2, \mathbf{v}_4, \mathbf{v}_6) \ .$$

Then the kinetic equations can be written as

$$\left[\frac{\partial}{\partial t} + c \cos (2i - 1)\frac{\pi}{6} \frac{\partial}{\partial x} + c \sin(2i - 1)\frac{\pi}{6} \frac{\partial}{\partial y} \right] N_i$$

$$= \frac{2}{3}cS(N_{i+1}N_{i+4} + N_{i+2}N_{i+5} - 2N_iN_{i+3}) \qquad (5.37)$$

$$+ \frac{5cS^{5/2}}{4\sqrt{\pi}}(N_{i+1}N_{i+3}N_{i+5} - N_iN_{i+2}N_{i+4}) \ .$$

In Section 2.1.4 a differently oriented model has been considered. Keeping this in mind, one can re-write the expressions of the Maxwellians and the relations between the parameters $\{A, c_x, c_y\}$ and the macroscopic variables $\{\hat{\nu}, \hat{u}_x, \hat{u}_y\}$. Since from (5.8) the y–component of the drift velocity at the wall must be equal to zero ($c_y = 0$), the Maxwellians in this case are re-written as

$$\hat{N}_1 = \hat{N}_6 = Ae^{c_x} \ , \quad \hat{N}_2 = \hat{N}_5 = A \ , \quad \hat{N}_3 = \hat{N}_4 = Ae^{-c_x} \ , \qquad (5.38)$$

with

$$c_x = \log \left[\frac{\frac{\hat{u}_x}{\sqrt{3}c} + \sqrt{1 - \frac{\hat{u}_x^2}{c^2}}}{1 - \frac{2\hat{u}_x}{\sqrt{3}c}} \right] \ , \quad 0 \le \hat{u}_x < \frac{\sqrt{3}}{2}c \ . \qquad (5.39)$$

Introducing the nondimensional quantities given in (5.26), Eq.(5.38) re-writes

$$\widehat{N}_1' = \widehat{N}_6' = \frac{e^{c_x}}{e^{c_x} + e^{-c_x} + 1}$$

$$\widehat{N}_2' = \widehat{N}_5' = \frac{1}{e^{c_x} + e^{-c_x} + 1} \qquad (5.40)$$

$$\widehat{N}_3' = \widehat{N}_4' = \frac{e^{-c_x}}{e^{c_x} + e^{-c_x} + 1} .$$

The evolution equations for the densities should be coupled to the boundary conditions (5.11). In this case a purely diffusive reflection law ($\alpha_D = 1$) is assumed, according to the fact that no specular reflection law can be applied on a moving boundary (see Section 5.1). On the upper plate also the x–component of the drift velocity of the re-emitted gas must vanish and therefore according to (5.12), (5.39) and (5.40) the re-emission rates are

$$\eta = 1 \quad : \quad B_4 = \frac{1}{4} \quad , \quad B_5 = \frac{1}{2} \quad , \quad B_6 = \frac{1}{4} . \qquad (5.41)$$

At the lower plate, the x-component of the mass velocity of the re-emitted gas must be equal to the wall velocity $w_w(t)$, therefore

$$\eta = 0 \quad : \quad B_1 = \frac{e^{c_x}}{e^{c_x} + e^{-c_x} + 2}$$

$$B_2 = \frac{2}{e^{c_x} + e^{-c_x} + 2} \qquad (5.42)$$

$$B_3 = \frac{e^{-c_x}}{e^{c_x} + e^{-c_x} + 2}$$

with

$$c_x = c_x(\tilde{w}_w) = \log \left[\frac{\frac{\tilde{w}_w}{\sqrt{3}} + \sqrt{1 - \tilde{w}_w^2}}{1 - \frac{2\tilde{w}_w}{\sqrt{3}}} \right] \quad , \quad 0 \leq \tilde{w}_w = \frac{w_w}{c} < \frac{\sqrt{3}}{2} . \quad (5.43)$$

Of course, here the parameter c has to be chosen so that $c > \frac{2}{\sqrt{3}} w_w$. The problem to be solved is then

$$\left[\frac{\partial}{\partial \tau} + \sin(2i-1)\frac{\pi}{6}\frac{\partial}{\partial \eta} \right] N_i' = \frac{2}{3 Kn_1} (N_{i+1}' N_{i+4}' + N_{i+2}' N_{i+5}' - 2 N_i' N_{i+3}')$$

$$+ \frac{1}{Kn_2} (N_{i+1}' N_{i+3}' N_{i+5}' - N_i' N_{i+2}' N_{i+4}') .$$

$$(5.44)$$

with boundary conditions

$$\eta = 0 \; : \; N_h'(\tau) = \frac{B_h}{\sin(2h-1)\frac{\pi}{6}} \sum_{k=4}^{6} | \sin(2k-1)\frac{\pi}{6} | \, N_k'(\tau) \; , \quad h = 1, 2, 3$$

$$(5.45)$$

$$\eta = 1 \; : \; N_k'(\tau) = \frac{B_k}{| \sin(2k-1)\frac{\pi}{6} |} \sum_{h=1}^{3} \sin(2h-1)\frac{\pi}{6} \, N_h'(\tau) \; , \quad k = 4, 5, 6$$

with B_h and B_k given by (5.41,5.42).

As an application consider the case in which the lower wall moves with dimensionless velocity

$$\tilde{w}_w = \frac{1}{2} \tanh \sigma\tau , \qquad (5.46)$$

and the initial conditions are

$$N_i'(\tau = 0, \eta) = \frac{1}{6} \qquad (5.47)$$

Since $\tilde{w}_w$ tends to $\frac{1}{2}$ as τ goes to ∞, a trend towards the steady state is expected. In fact all the figures show, after a transient behaviour, such a tendency. Furthermore, similarly to the thermal creep problem, the x–component of the drift velocity at the wall is not equal to the wall velocity, as it is also known from experimental result. It is possible, therefore, to define the slip coefficient

$$\Delta_u = w_w(\tau) - u_x(\eta = 0, \tau) , \qquad (5.48)$$

which is a parameter of physical interest.

The solution of the initial-boundary value problem is obtained applying the numerical technique used in the previous section and leads to the results indicated in Figs. 5.8–5.11 for $Kn_1 = 0.1$ and $Kn_2 = 10^4$. In Figs. 5.9–5.11 σ in (5.46) is set equal to one.

In particular, in Fig. 5.8 the wall velocity, the x–component of the mass velocity on the moving wall and the slip coefficient at the moving wall are plotted versus time, for $\sigma = 1$ and $\sigma = 8$. Note, in particular, the presence of a peculiar overshoot of the slip coefficient in response to a large acceleration of the wall and the fast trend of the plotted quantities towards the steady state.

In Fig. 5.9 the components u_x and u_y of the mean velocity are represented for different values of $\eta \in [0, 1]$ versus time. In Fig. 5.10 the y–component of the mass velocity in the center of the channel $\eta = \frac{1}{2}$ is

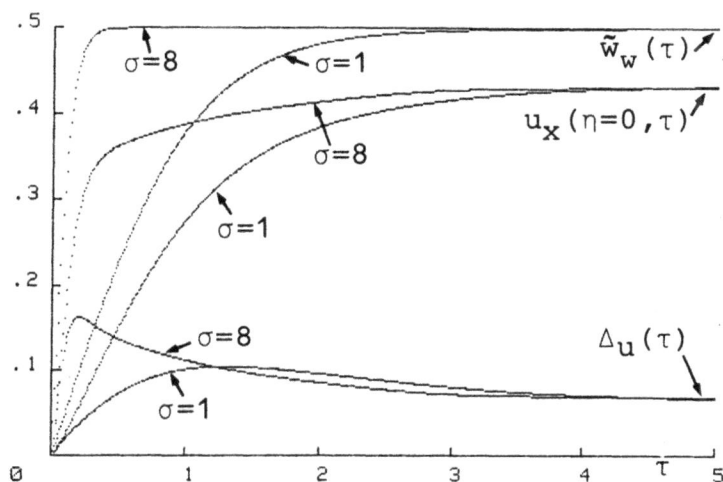

Fig. 5.8 — *Wall velocity, tangential drift velocity near the wall and slip coefficient versus times, for different wall velocities*

more deeply examined on a wider range of time to show the propagation of a damped wave which goes back and forth between the plates and tends to zero asymptotically. In fact from the *steady* kinetic equations of the models it readily follows that $u_y = 0$ everywhere.

Finally in Fig. 5.11 the time-space behaviour of the total numerical densities $\nu(\tau, \eta)$ is shown for different locations.

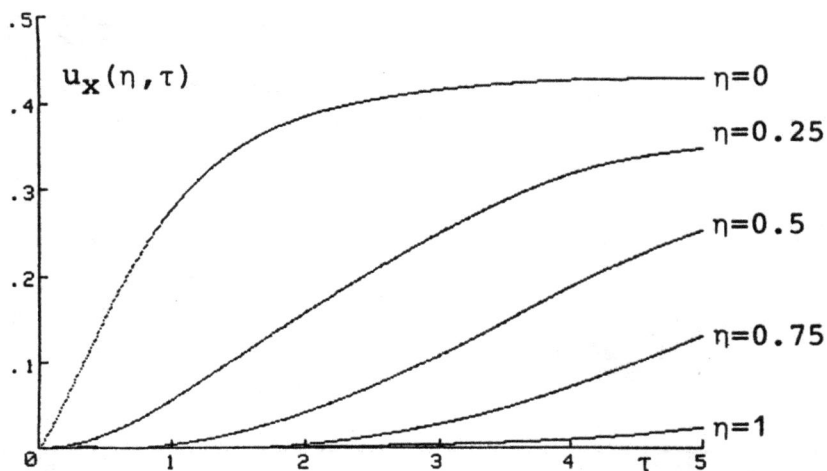

(a)

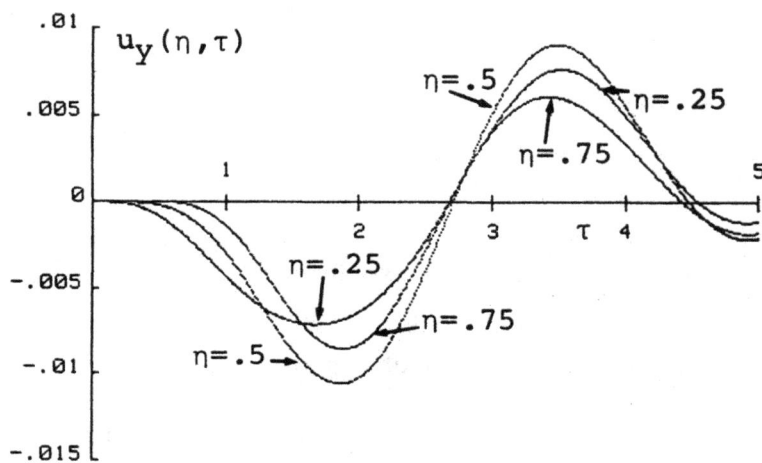

(b)

Fig. 5.9 — *Distibution of the (a) tangential and (b) normal drift velocity versus time in different locations between the plates*

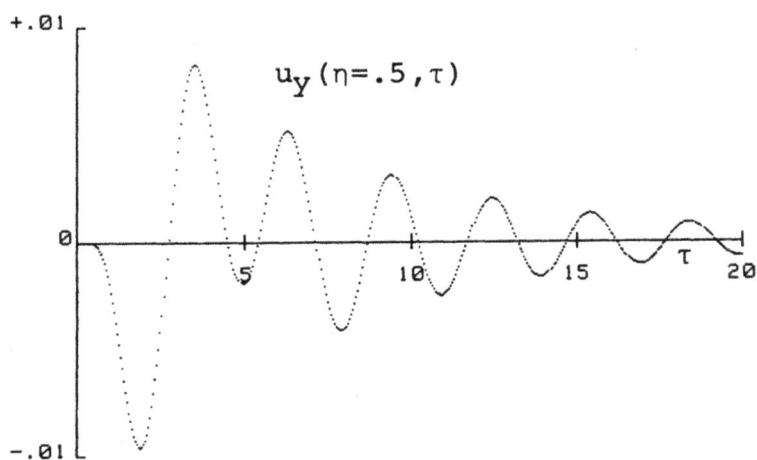

Fig. 5.10 — *Normal drift velocity in the core of the flow vs time*

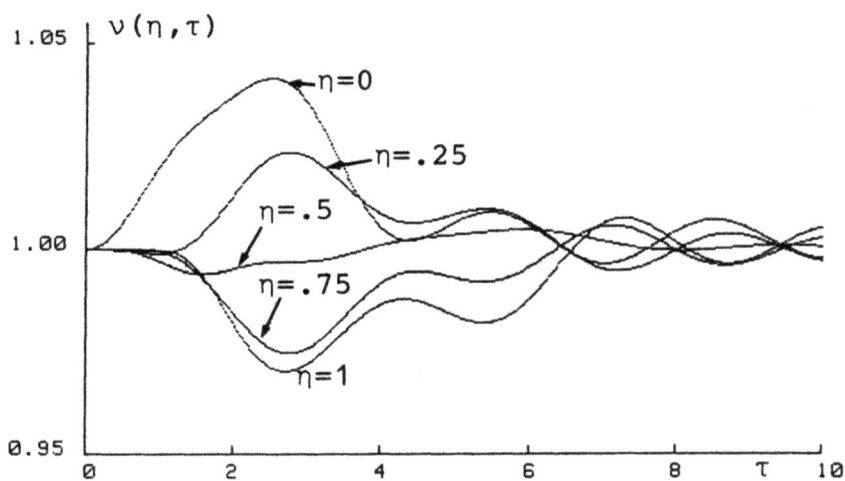

Fig. 5.11 — *Numerical density vs time in different locations*

5.5   Rayleigh Flow

Also this type of flow, as those described in the preceding sections, is a classical one in fluid-dynamics and is somewhat similar to Couette flow.

In this present case, the gas is confined in a half-space $y > 0$, by a solid wall located at $y = 0$. At $t = 0$, the gas is at rest with initial conditions (5.34) and $w_w = 0$. For $t > 0$, the wall starts moving with a constant (or time-dependent) speed. Once more the initial-boundary value problem needs to be solved to obtain the densities $N_i(t, y)$ which allow to recover the macroscopic observables. The conditions at infinity are equal to the initial conditions.

Such a problem was solved analytically by Broadwell [14] for his model and can be dealt with in the same fashion by the 4–velocity model for a binary mixture. Detailed calculation can be found in ref.[17]. The analytical expressions of the solution and the main steps of the calculations will be given here.

Consider Eqs.(2.9) with $\frac{\partial N_i^1}{\partial x} = \frac{\partial N_i^2}{\partial x} = 0$, with e_i given by

$$e_1 = -e_3 = \frac{1}{\sqrt{2}}(\mathbf{i} + \mathbf{j}) \;,\quad e_2 = -e_4 = \frac{1}{\sqrt{2}}(-\mathbf{i} + \mathbf{j}) \;,\qquad (5.49)$$

with initial data

$$\widehat{N}_i^1(t = 0, y) = N_o^1 = \text{const.}$$

$$\widehat{N}_i^2(t = 0, y) = N_o^2 = \text{const.}$$

and with the boundary conditions on $y = 0$ given by (5.11) with $\alpha_D = 1$. The bottom plate moves instantaneously for $t > 0$ with constant speed w_w directed in the positive direction of the x–axis. The number densities can be rescaled as follows

$$\tilde{N}_i^r = \frac{N_i^r}{N_o^1} \quad , \quad r = 1, 2 \ .$$

Exploiting symmetry properties of the kinetic equations (see ref.[17] for details) one finds that

$$\tilde{N}_1^r = 2\psi - \tilde{N}_2^r \ , \quad \tilde{N}_4^r = 2\psi - \tilde{N}_3^r \ , \quad r = 1, 2 \tag{5.50}$$

with $\psi = \dfrac{N_o^2}{N_o^1}$.

Introducing the new dimensionless variables

$$\tau = 2c\tilde{S}N_o^1 t \quad , \quad \eta = 2\sqrt{2}\tilde{S}N_o^1 y \ ,$$

where

$$\tilde{S} = S_{11} + \frac{1}{2}S_{12}(1 + \mu)\psi^2 \ ,$$

the initial-boundary value problem is reduced to the equation

$$\frac{\partial^2 \tilde{N}_2^1}{\partial \tau^2} - \frac{\partial^2 \tilde{N}_2^1}{\partial \eta^2} + \frac{\partial \tilde{N}_2^1}{\partial \tau} = 0 \tag{5.51}$$

with initial data

$$\tilde{N}_2^1(\tau = 0, \eta) = 1$$

and boundary conditions

$$\tilde{N}_2^1(\tau, \eta = 0) = 1 - w_w' \quad , \quad w_w' = \sqrt{2}\,\frac{1 + \psi}{1 + \mu\psi}\,\frac{w_w}{c} \; .$$

Using the symmetry conditions yields

$$\tilde{N}_3^1 = \tilde{N}_2^1 + 2\left(\frac{\partial \tilde{N}_2^1}{\partial \eta} + \frac{\partial \tilde{N}_2^1}{\partial \tau}\right) \; . \tag{5.52}$$

Equation (5.51) can be solved analytically (see, for instance, ref.[22], vol.2, Chap.7), with result

$$\tilde{N}_2^1(\tau, \eta) = 1 - w_w' H(\tau - \eta)\left[e^{-\eta/2} + \frac{\eta}{4}\int_\eta^\tau e^{-z/2}\,\frac{I_1(\zeta)}{\zeta}\,dz\right]$$

where $\zeta = \frac{1}{2}(z^2 - \eta^2)^{1/2}$, $H(\tau - \eta)$ is the Heaviside function

$$H(\tau - \eta) = \begin{cases} 0 & \text{if } \tau \le \eta \\ 1 & \text{if } \tau > \eta \end{cases}$$

and I_1 is the modified Bessel function of first kind and first order.

Finally all the number densities can be recovered by Eqs.(5.50) and (5.52).

Even if this application refers to a very simple model (only four velocities per gas component), the application itself has still been summarized. In fact it refers to one of the few cases in which an analytic solution is available. The reader can develop calculations, say by the differential quadrature method or by finite difference schemes, for more general models and test the efficiency of the numerical scheme using this analytic solution. This can be regarded as a suggested exercise.

5.6 Flow over Convex Bodies

The analysis of the flow patterns over convex bodies is certainly of relevant interest for the applications in aerodynamics. In fact one expects to obtain from this type of computing the calculations of the aero-thermodynamical quantities: drag, lift and heat transfer coefficients.

An example of this type of calculations obtained by the full Boltzmann equation is the one of paper [8], which shows, however, how the difficulty in dealing with the nonlinear Boltzmann equation obliges one to use somewhat heuristic approaches to the problem, motivated by the need of obtaining engineering results.

On the other hand, the discrete Boltzmann equation is certainly a valid candidate to solve such a problem. In fact its numerical treatment is, as we have already seen, an obstacle not too hard to be dealt with.

No result in the literature is known to deal with the analysis of the flow over convex bodies by the discrete Boltzmann equation. Therefore, the mathematical formulation of such a problem will be provided in this section also with the aim of giving suitable challenging suggestions to applied mathematicians.

Keeping this in mind, consider a two-dimensional convex body Ω and a frame defined by a rectangle with border located at a *sufficiently large* distance from the body, as shown in Fig. 5.12. The meaning of sufficiently large will be commented afterwards.

The analysis of the problem consists in solving the boundary value problem for the steady discrete Boltzmann equation

$$\mathbf{v}_i \cdot \nabla_{\mathbf{x}} N_i = J_i[\mathbf{N}] \qquad (5.53)$$

with the boundary conditions defined on the body and on the border of the rectangle. On the border one assumes N_i to be defined by Maxwellian distributions with given mass velocity

$$\mathbf{u}_\infty = u_\infty \mathbf{i} \ .$$

Moreover on the body, the surface of which will be indicated by $\partial\Omega$, the boundary conditions are given assigning

$$\mathbf{x} \in \partial\Omega \quad : \quad B_{ij} \text{ for } i \in I_i \text{ and } j \in I_r \ . \qquad (5.54)$$

In particular B_{ij} can be defined in terms of Maxwellian boundary conditions as in (5.21).

In order to obtain quantitative results, it may be convenient to deal with a time dependent technique based upon the formulation of an initial-boundary value problem.

In this case Eq.(5.1), instead of Eq.(5.53), will be considered with initial conditions set equal to suitable Maxwellians, namely

$$N_{io} = N_i(t = 0, \mathbf{x}) = \widehat{N}_i \qquad (5.55)$$

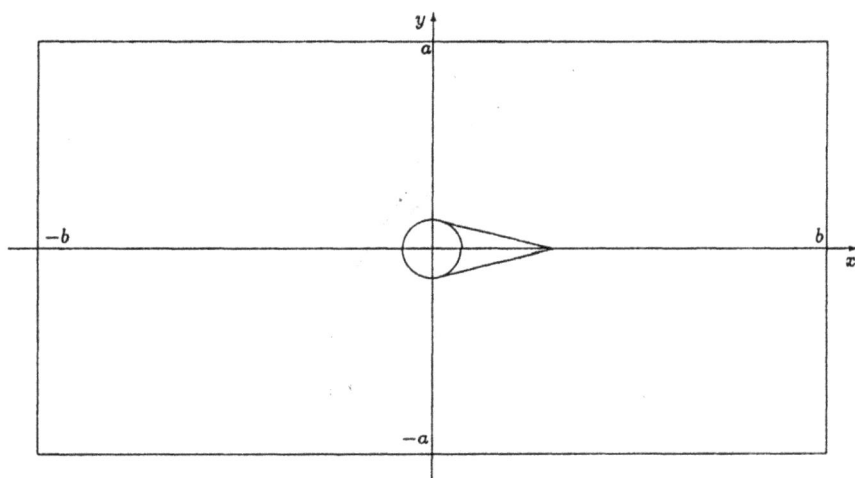

Fig. 5.12 — *Geometry of the system*

and with the boundary conditions which have been stated above. Then one should investigate whether a steady flow is reached asymptotically in time.

The final aim of the analysis which has been here presented is to compute the drag $\mathcal{D}$, the lift forces $\mathcal{L}$ and the heat transfer $\mathcal{H}$ over the body. In order to show the technical computing of these quantities, consider a surface element $d\Omega$ with outer normal n as indicated in Fig. 5.13.

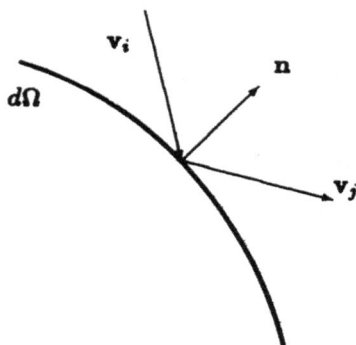

Fig. 5.13 — *Surface element of $\partial\Omega$*

The incoming and outgoing momentum fluxes on the unit surface $d\Omega$, respectively $\mathbf{Q}_i$ and $\mathbf{Q}_r$, are given by

$$
\begin{aligned}
\mathbf{Q}_i &= \sum_{i \in I_i} |\mathbf{v}_i \cdot \mathbf{n}| \mathbf{v}_i N_i \\
\mathbf{Q}_r &= \sum_{j \in I_r} |\mathbf{v}_j \cdot \mathbf{n}| \mathbf{v}_j N_j \ .
\end{aligned}
\tag{5.56}
$$

In the same fashion one can compute the energy fluxes $\mathcal{E}_i$ and $\mathcal{E}_r$, respectively

$$
\begin{aligned}
\mathcal{E}_i &= \sum_{i \in I_i} \frac{1}{2} m v_i^2 |\mathbf{v}_i \cdot \mathbf{n}| N_i \\
\mathcal{E}_r &= \sum_{j \in I_r} \frac{1}{2} m v_j^2 |\mathbf{v}_j \cdot \mathbf{n}| N_j \ .
\end{aligned}
\tag{5.57}
$$

Considering that the lift and drag forces are the resultant of the momentum variation in the direction orthogonal and parallel to $\mathbf{u}_\infty$, respectively, one has

$$\mathcal{L} = m \int_\Omega (\mathbf{Q}_i - \mathbf{Q}_r) \cdot \mathbf{j} \, d\Omega$$

$$\mathcal{D} = m \int_\Omega (\mathbf{Q}_i - \mathbf{Q}_r) \cdot \mathbf{i} \, d\Omega \, .$$

(5.58)

Analogous is the calculation of the heat transfer over the surface

$$\mathcal{H} = \int_\Omega (\mathcal{E}_i - \mathcal{E}_r) \, d\Omega \, .$$

(5.59)

Classically [1,2] the quantities which have been defined above, are given in terms of dimensionless coefficients, i.e. the drag, lift and heat transfer coefficients, respectively

$$C_D = \frac{\mathcal{D}}{\frac{1}{2}\rho u_\infty^2 S_\Omega}$$

$$C_L = \frac{\mathcal{L}}{\frac{1}{2}\rho u_\infty^2 S_\Omega}$$

$$C_H = \frac{\mathcal{H}}{\frac{1}{2}\rho u_\infty^3 S_\Omega}$$

(5.60)

where S_Ω is the cross sectional area of the body and ρ is the mass density of the gas. Of course a crucial aspect in the analysis of such a problem is the correct choice of the dimensions a and b of the rectangular frame around the convex body. The dimensions of the frame need to be of a higher order of magnitude compared with the dimensions of the body and to the one of the mean free path. In this case a sufficient number of collisions will occur within the frame in order to allow the gas to reach the asymptotic conditions at *infinity*. In other words, the frame must be sufficiently large to allow the simulation of the behaviour of the gas at infinity in space.

References

[1] M. Kogan, **Rarefied Gas Dynamics**, Plenum Press, New York, 1969.

[2] A.H. Shapiro, **The Dynamics and Thermodynamics of Compressible Fluid Flow**, Ronald Press, New York, 1954.

[3] R. Gatignol, "Kinetic boundary conditions for discrete theory of gases", *Phys. Fluids*, **20**, 1977, p.2022.

[4] L. Preziosi and E. Longo, "On the decomposition of domains in nonlinear discrete kinetic theory", in **Discrete Models of Fluid Dynamics**, *Advances in Mathematics for Applied Sciences* vol.2, Ed. A. Alves, World Scientific, Singapore, London, 1991, p.144.

[5] P. Resibois and M. de Leener, **Classical Kinetic Theory of Fluids**, Wiley, London, 1977.

[6] C. Cercignani, **Theory and Application of the Boltzmann Equation**, Scottish Academic Press, New York, 1975.

[7] N. Bellomo, C. Dankert, H. Legge and R. Monaco, "Drag, heat flux and recovery factor in free molecular flow and gas surface interaction analysis", in **Rarefied Gas Dynamics**, Eds. O.M. Belotserkovski, M.N. Kogan, S.S. Kutateladze and A.K. Rebrov, Plenum Press, New York, 1985, p.421.

[8] N. Bellomo, N. de Divitiis and L.M. de Socio, "Fluid-dynamic forces on a tethered satellite system", in **Rarefied Gas Dynamics**, Ed. A. Beylich, VCH-Verlag, Weinheim, New York, 1991, p.369.

[9] S. Kawashima, "Initial-boundary value problem for the discrete Boltzmann equation", in **Equationes aux derivees partielle**, Centre Math., Ecole Politecnique, Palaiseau (France), Octobre 1988, III-1-6.

[10] S. Kawashima, "Global existence and asymptotic behaviour of solutions to the mixed problem for the discrete Boltzmann equation", *Nonlinear Anal.*, to appear in 1991.

[11] Z. Brzezniak, F. Flandoli and L. Preziosi, "On the discrete Boltzmann equation with multiple collisions", *Stab. and Appl. Anal. in Continuous Media*, to appear in 1991.

[12] A. Pulvirenti, "Global solutions to the initial–boundary value problem for the discrete Boltzmann equation", *Transp. Theory Statist. Phys.*, to appear in 1991.

[13] C. Cercignani, R. Illner and M. Shinbrot, "A boundary value problem for the two dimensional Broadwell model", *Comm. Math. Phys.*, **114**, 1988, p.687.

[14] J. Broadwell, "Study of rarefied shear flow by the discrete velocity method", *J. Fluid Mech.*, **19**, 1964, p.401.

CHAPTER 6

THE DISCRETE BOLTZMANN EQUATION

FOR CHEMICALLY REACTING GASES

6.1 Gases with Chemical Reactions

Mathematical models for fluid-dynamics of chemically reacting gases are certainly of great interest for the applications. This statement
holds true also in the case of the kinetic theory of gases. In fact chemical
reactions can occur in several physically interesting situations of rarefied
gas dynamics, say fluid-dynamics of ionized gases, gases which reacts
chemically because of the impact on obstacles, combustion theory and
so on.

Examples of reactions between gases can be found in the book by
Bird [1] who treated in a systematic way the kinetic theory of chemically
reacting gases. Moreover other examples in combustion theory and

modelling are indicated in the survey paper by Peters and Kennel [2].

Some of the chemical reactions occurring in nature can be framed by the following schemes

a) $A + B \longrightarrow C + D$

b) $A + B \longrightarrow C$

$A + A \longrightarrow B$

c) $A + B + C \longrightarrow C + D$.

Reactions of type (a) are generally called bi-molecular, while class (b) is sometimes referred to as dissociation-recombination reactions. Finally class (c) is indicated in the literature as catalytic or tri-molecular reactions. Types (a) and (c) are often also reversible. Examples are, respectively

a) $CO + OH \rightleftharpoons CO_2 + H$

$CH_4 + H \longrightarrow CH_3 + H_2$

b) $C_5 H_{12} \longrightarrow C_4 H_9 + CH_3$

$I_2 + I_2 \Big\langle \begin{array}{c} 2I + I_2 \\ 4I \end{array} \quad , \quad I + I = I_2$

c) $CH_3 + H + M \longrightarrow CH_4 + M$

$HCO + M \longrightarrow CO + H + M$

M being a suitable catalyzer.

Other examples of dissociation-recombination reactions can be found

in the cycle of ozone

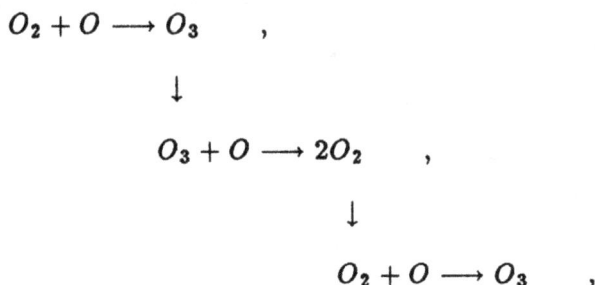

$$O_2 + O \longrightarrow O_3 \quad ,$$

$$\downarrow$$

$$O_3 + O \longrightarrow 2O_2 \quad ,$$

$$\downarrow$$

$$O_2 + O \longrightarrow O_3 \quad ,$$

when oxygen molecules collide with their atoms which have been pro-duced by ultra-violet radiation.

Bi-molecular reactions are, conversely, typical in the chain of ozone destruction due to the presence of cloride atoms from fluoro-chlorocarbons in upper atmosphere,

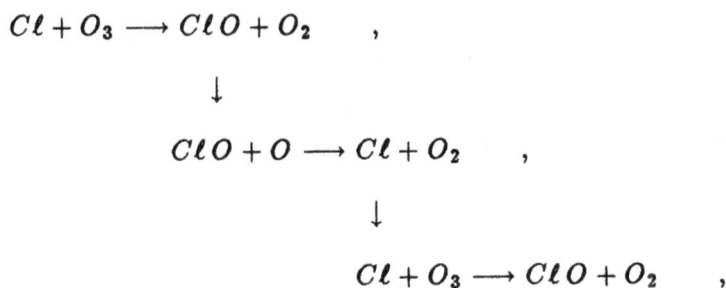

$$Cl + O_3 \longrightarrow ClO + O_2 \quad ,$$

$$\downarrow$$

$$ClO + O \longrightarrow Cl + O_2 \quad ,$$

$$\downarrow$$

$$Cl + O_3 \longrightarrow ClO + O_2 \quad ,$$

Some results on chemically reacting gases in kinetic theory are given in [3–6]. These papers refer to the Boltzmann equation for gas mixtures or to the Boltzmann equation in the framework of the scattering kernel formulation. Even for the Enskog gas some results are known in literature [7,8]. Also in lattice gas theory the problem of simulation of reacting collisions has been dealt with (see, for instance, ref.[9]).

This chapter reports about some preliminary research developed towards a discrete kinetic theory for chemically reacting gases with reference to reactions of type (a) and (b). We may add some further motivations to justify the study of reacting gases by the discrete theory. In fact the full Boltzmann equation for a gas with chemical reactions shows a higher complexity with respect to the discrete kinetic theory so that simplified models are useful. Furthermore, chemical reactions may need, as in the case of reaction of type (c), multiple collisions which cannot be dealt with by the full Boltzmann equation, but can be described, as we have seen starting from Chapter 1, by the discrete Boltzmann equation.

The content of this chapter is somewhat different from that of the preceding ones. In fact the discrete kinetic theory for neutral gases is based upon well established theories; on the contrary very little is known about the discrete kinetic theory with chemical reactions and all is essentially contained in three very recent papers [10–12]. Therefore, the chapter will report their contents and provide suitable indications on several still open reasearch areas.

The chapter is organized in six sections. The second section, after this brief introduction, deals with the modelling of the discrete Boltzmann equation for bi-molecular reactions. The third section studies a suitable planar model in which both dissociations of molecules into atoms and recombinations of atoms in molecules are allowed. The fourth section deals with the modelling of the discrete Boltzmann equation for a dissociating gas, which is then applied in the fifth section to study the propagation of a shock wave. Finally the last section provides a discussion on several open problems giving the reader several research suggestions.

It is plain that the modelling aspects will strongly relay on the theory of the discrete Boltzmann equation for gas mixtures which has been dealt with in the first two chapters of this book.

6.2   A Mathematical Model for Gases with Bi-Molecular Chemical Reactions

This section presents, with reference to the analysis of paper [11], a mathematical model, in the framework of the discrete Boltzmann equation, for chemically reacting gases, referred to a gas of four species undergoing simple elastic binary collisions and so-called *reversible bimolecular* reactions, which may be schematically written as

$$A_1 + A_2 \rightleftharpoons A_3 + A_4$$

where the rate of formation of each component will depend on the concentration of the various species described by suitable *reaction rate* coefficients to be modelled in terms of functions of the temperature (ref.[1], Chap.XII).

Consider a gas with four species each with molecular mass m_p, $p = 1, \ldots, 4$, such that $\mu_p = m_1 / m_p \leq 1$ is the ratio referred to the lightest gas of mass m_1 . In addition let $m_1 + m_2 = m_3 + m_4$, in order to assure mass conservation in the chemical reactions.

Assume a velocity discretization with 24 velocities, six for each gas, such that

$$i = 1, \ldots, 6, \quad p = 1, \ldots, 4: \qquad \mathbf{v}_i^p = \mu_p c \mathbf{e}_i$$

where $\mathbf{e}_i$, $i = 1, 2, 3$ are the unit vectors directed along the coordinate axis x, y and z, respectively and

$$\mathbf{e}_i = -\mathbf{e}_{i-3} \text{ for } i = 4, 5, 6 .$$

The model is then a 4×6–velocity Broadwell model for a mixture of four gas-components (see Section 2.2.2).

The number densities $N_i^p = N_i^p(t, \mathbf{x}) : [0, \mathcal{T}] \times \mathbf{R}^3 \longmapsto \mathbf{R}_+$ are joined to each velocity (i) and to each species (p). Then, in the framework of discrete kinetic theory, the mathematical modelling of the physical system, which has been described above, consists in deriving suitable evolution equations for the densities N_i^p when the gas-particles undergo bi-molecular interactions together with the usual mechanical collisions.

Keeping this in mind it is possible to state, still at a quite general level, the following assumptions:

i) Only simple binary collisions in absence of external force fields preserving mass, momentum and energy are taken into account (here conservation of energy means conservation of the sum of mechanical and chemical energies). Interaction between pairs of particles of different gases can occur. The gas molecules are assumed to be polyatomic but the effects of vibrational and rotational degrees of freedom will be neglected.

ii) If an interaction occurs between particles of the first and second gases, i.e. for $p = 1$ and $p = 2$, then a fraction K_1 of particles is transferred into the third and fourth gases, whereas the fraction $(1 - K_1)$ remains in the same gases. The analogous situation is assumed for collisions between the third and fourth gases, where only a fraction K_2 of particles is transferred into the first and second gas. Then a fraction $(1 - K_2)$ of the third and fourth gas undergo elastic interactions, remaining in the same gases.

iii) Collisions between a particle of the first two gases and one of the second two gases do not give rise to chemical reactions, but only to elastic interactions.

iv) The coefficients $K_1 \leq 1$ and $K_2 \leq 1$ are assumed to be given functions of the local temperature T of the gas.

The modelling of the reaction rates K_1 and K_2, which can be considered as chemical affinity coefficients, is well established in the literature (see ref.[1], Chap.XII). In particular it can be assumed that K_1 and K_2 are defined by the following laws

$$
\begin{aligned}
K_1 &= r_1 T^\xi e^{-s_1/T} \\
K_2 &= r_2 T^\xi e^{-s_2/T} \quad,
\end{aligned}
\tag{6.1}
$$

where r_1, r_2, s_1, s_2 and ξ are constants which may depend upon the chemical properties of the four gas species. In the particular case $\xi = 0$, the aforementioned law is known as the Arrhenius model. An example will be given in Section 6.5.

The hypothesis of item (i) contains a simplification which is necessary in the framework of discrete velocity models, that is the one of monoatomic gases. Nevertheless, according to the already quoted book [1], bi-molecular reactions can be modelled, at least in a somewhat approximated way, without considering internal degrees of freedom.

The derivation of the evolution equations for the densities can be realized resorting to ideas analogous to the ones used in the discrete kinetic theory of non reacting gases, namely by equating the total derivative of N_i^p to the corresponding gain and loss terms due to both reacting and non reacting collisions

$$\frac{\partial N_i^p}{\partial t} + \mathbf{v}_i^p \cdot \nabla_\mathbf{x} N_i^p = G_i^p[\mathbf{N}] - L_i^p[\mathbf{N}] \tag{6.2}$$

where $\mathbf{N} = \{N_i^p\} \in \mathbf{R}^{24}, i = 1, \ldots, 6$ and $p = 1, \ldots, 4$.

A detailed calculation of the terms G_i^p and L_i^p can now be derived considering each type of collisions separately. Referring to collisions without chemical reactions, that is when $K_1 = K_2 = 0$, between particles with velocities $\mathbf{v}_i^p$ and $\mathbf{v}_j^r$, one has, as usual,

$$p, r = 1, \ldots, 4 \qquad i, j, h, k = 1, \ldots, 6$$

$$G_i^p = A_{ph,rk}^{pi,rj} N_h^p N_k^r \quad , \quad L_i^p = A_{pi,rj}^{ph,rk} N_i^p N_j^r \tag{6.3}$$

where the transition rates A, which give the probability that a p-particle of i-velocity colliding with a r-particle with j-velocity will have after collision, respectively, h and k–velocities, are related to the transition probability densities a by

$$A^{ph,rk}_{pi,rj} = S_{pr}|\mathbf{v}^p_i - \mathbf{v}^r_j|a^{ph,rk}_{pi,rj}$$

where the term S_{pr} denotes the pair cross sectional area and $a^{ph,rk}_{pi,rj}$ satisfy the microreversibility and indistinguishability properties (1.23). This kind of modelling has already been studied in Section 1.2.2.

When chemical reactions occur, that is $K_1 \neq 0$, $K_2 \neq 0$, the expressions of the gain and loss terms are still given as in Eqs.(6.3), but only indistinguishability holds, that is

$$a^{ph,rk}_{pi,rj} = a^{ph,rk}_{rj,pi} = a^{rk,ph}_{pi,rj} = a^{rk,ph}_{rj,pi} \; ; \; a^{ph,rk}_{pi,rj} \neq a^{pi,rj}_{ph,rk} . \tag{6.4}$$

Reversibility property can be applied only to purely mechanical collisions.

6.2.1 Kinetic Equations of the Discrete Model with Bi-Molecular Chemical Reactions

After the preliminaries of the preceding section it is possible to consider, separately, all classes of non reacting and reacting collisions and the evolution of densities.

—   *Collisions between Particles of the Same Gas*

This type of collision corresponds to the ones of the model proposed by Broadwell for a simple gas. The only collisions which generate some changes in the fluxes are the

i) Head-on collisions between particles of the same gas with relative velocity $2\mu_p c$, probability density $\frac{1}{3}$ and scattering in the directions at right angles with respect to pre-collisional velocities.

The corresponding gain and loss terms are respectively given by the following expressions,

$$G_i^{pp} = \frac{2}{3}\mu_p c S_{pp}(N_{i+1}^p N_{i+4}^p + N_{i+2}^p N_{i+5}^p)$$
$$L_i^{pp} = \frac{4}{3}\mu_p c S_{pp} N_i^p N_{i+3}^p \qquad\qquad (6.5)$$

for $i = 1, \ldots, 6$, where $i+n > 6 \Longrightarrow i+n = i+n-6$. These expressions have already been given in Chapter 2 and are reported here, in the new notations, for sake of completeness.

—   *Collisions without Chemical Reactions between Particles of Different Gases*

This type of collision corresponds to the ones of the Broadwell model for gas mixtures already studied in Section 2.2.2. The only difference is now that one has to take into account that only the fractions $(1 - K_1)$ and $(1 - K_2)$ of the interactions between the species

$$(p = 1, \ r = 2) \quad \text{and} \quad (p = 3, \ r = 4) \ ,$$

respectively, undergo elastic collisions without chemical reactions.

Obviously interactions of types

$$(p = 1, \; r = 3) \; , \quad (p = 1, \; r = 4) \; , \quad (p = 2, \; r = 3) \; , \quad (p = 2, \; r = 4)$$

do not produce chemical reactions and must be taken into account within this kind of collisions. The collisions which will give rise to change in the fluxes are the following

ii) Head-on collisions with relative velocity $c(\mu_p + \mu_r)$, products of the type

$$i = 1, \ldots, 6 \; : \quad (N_i^p, N_{i+3}^r) \longleftrightarrow \begin{cases} (N_{i+1}^p, N_{i+4}^r) \\[2mm] (N_{i+4}^p, N_{i+1}^r) \\[2mm] (N_{i+2}^p, N_{i+5}^r) \\[2mm] (N_{i+5}^p, N_{i+2}^r) \\[2mm] (N_{i+3}^p, N_i^r) \end{cases}$$

and probability densitiy $\frac{1}{6}\gamma_{pr}$ where γ_{pr} is equal to $(1-K_1), (1-K_2)$ or 1 according to whether the particles may react or not.

iii) Scattering of particles colliding at $90°$ with relative velocity $c(\mu_p^2 + \mu_r^2)^{\frac{1}{2}}$, probability density $\frac{1}{2}\gamma_{pr}$ and products

$$i, j = 1, \ldots, 6 \; , \quad j \neq i, i+3 \; : \quad (N_i^p, N_j^r) \longleftrightarrow (N_j^p, N_i^r) \; .$$

The corresponding gain and loss terms for collisions (ii) and (iii) are then

$$G_i^{pr} = c \sum_{r \neq p} \gamma_{pr} S_{pr} [\frac{\mu_p + \mu_r}{6}$$

$$\cdot (N_i^r N_{i+3}^p + N_{i+1}^p N_{i+4}^r + N_{i+4}^p N_{i+1}^r + N_{i+2}^p N_{i+5}^r + N_{i+5}^p N_{i+2}^r)$$

$$+ \frac{(\mu_p^2 + \mu_r^2)^{\frac{1}{2}}}{2} N_i^r (N_{i+1}^p + N_{i+2}^p + N_{i+4}^p + N_{i+5}^p)]$$

$$(6.6)$$

$$L_i^{pr} = c \sum_{r \neq p} \gamma_{pr} S_{pr} [5 \frac{\mu_p + \mu_r}{6} N_i^p N_{i+3}^r +$$

$$+ \frac{(\mu_p^2 + \mu_r^2)^{\frac{1}{2}}}{2} N_i^p (N_{i+1}^r + N_{i+2}^r + N_{i+4}^r + N_{i+5}^r)] ,$$

$$(6.7)$$

where

$$\gamma_{12} = \gamma_{21} = 1 - K_1 \quad , \quad \gamma_{34} = \gamma_{43} = 1 - K_2$$

$$\gamma_{13} = \gamma_{31} = \gamma_{14} = \gamma_{41} = \gamma_{23} = \gamma_{32} = \gamma_{24} = \gamma_{42} = 1 .$$

$$(6.8)$$

— *Collisions with Chemical Reactions*

This type of interaction occurs, with a fraction K_1, between two particles of the first two gases which are then transferred into the second two gases, and vice versa with fraction K_2.

The collisions which have to be taken into account are the following:

iv) Head-on collisions of couples of particles of the first two gases with fraction K_1 (or of the second two gases with fraction K_2), with relative velocity $c(1 + \mu_2)$ (or $c(\mu_3 + \mu_4)$, respectively), with probability density $\frac{1}{6}\delta_{pr}$ and products along the same directions of collisions of type ii). Here again δ_{pr} is equal to K_1, K_2 or 0 according to whether the particles may react or not.

v) Scattering of particles colliding at $90°$ of the first two gases with fraction K_1 (of the second two gases with fraction K_2), with relative velocity $c(1+\mu_2^2)^{\frac{1}{2}}$ (with relative velocity $c(\mu_3^2+\mu_4^2)^{\frac{1}{2}}$, respectively), with probability density $\frac{1}{2}\delta_{pr}$ and products along the same directions of collisions of type (iii).

The corresponding gain and loss terms are analogous to the ones of Eqs.(6.6–6.7) with the differences indicated below

$$
\begin{aligned}
\mathcal{G}_i^{pr} = c \sum_{r \neq p} &\delta_{(p+2)(r+2)} S_{(p+2)(r+2)} \\
\cdot \{ &\frac{\mu_{p+2} + \mu_{r+2}}{6} (N_i^{p+2} N_{i+3}^{r+2} + N_i^{r+2} N_{i+3}^{p+2} + N_{i+1}^{p+2} N_{i+4}^{r+2} \\
&+ N_{i+4}^{p+2} N_{i+1}^{r+2} + N_{i+2}^{p+2} N_{i+5}^{r+2} + N_{i+5}^{p+2} N_{i+2}^{r+2}) \\
+ &\frac{(\mu_{p+2}^2 + \mu_{r+2}^2)^{\frac{1}{2}}}{2} [N_i^{r+2}(N_{i+1}^{p+2} + N_{i+2}^{p+2} + N_{i+4}^{p+2} + N_{i+5}^{p+2}) \\
+ &N_i^{p+2}(N_{i+1}^{r+2} + N_{i+2}^{r+2} + N_{i+4}^{r+2} + N_{i+5}^{r+2})]\}
\end{aligned}
\tag{6.9}
$$

$$
\begin{aligned}
\mathcal{L}_i^{pr} = c \sum_{r \neq p} &\delta_{pr} S_{pr}[(\mu_p + \mu_r) N_i^p N_{i+3}^r \\
+ &(\mu_p^2 + \mu_r^2)^{\frac{1}{2}} N_i^p (N_{i+1}^r + N_{i+2}^r + N_{i+4}^r + N_{i+5}^r)]
\end{aligned}
\tag{6.10}
$$

where

$$p + 2 > 4 \Longrightarrow p + 2 = p + 2 - 4$$

$$r + 2 > 4 \Longrightarrow r + 2 = r + 2 - 4$$

$$\delta_{12} = \delta_{21} = K_1 \quad , \quad \delta_{34} = \delta_{43} = K_2 \tag{6.11}$$

$$\delta_{13} = \delta_{31} = \delta_{14} = \delta_{41} = \delta_{23} = \delta_{32} = \delta_{24} = \delta_{42} = 0 \ .$$

Substituting all expressions (6.5–6.11) into (6.2) provides the final form of the mathematical model

$$\frac{\partial N_i^p}{\partial t} + \mathbf{v}_i^p \cdot \nabla_{\mathbf{x}} N_i^p = G_i^{pp} + G_i^{pr} + \mathcal{G}_i^{pr} - (L_i^{pp} + L_i^{pr} + \mathcal{L}_i^{pr}) \tag{6.12}$$

for $i = 1, \ldots, 6$ and $p, r = 1, \ldots, 4$.

6.2.2 Thermodynamic Equilibrium in Presence of Bi-Molecular Reactions

It is now possible to discuss some properties of the thermodynamic equilibrium.

The first step is the identification of the parameter c in the proposed velocity discretization. The collisional scheme (i)–(v) assumed in the previous section preserves mass and momentum during collisions. On the other hand, energy conservation is assured automatically for collisions (i)–(iii), which refer to neutral collisions, thanks to the choice of the velocity moduli equal to $\mu_p c$ for each species p. In the collisions (iv)–(v), with the presence of chemical reactions, the difference of kinetic energy of particles before and after collisions is given by

$$m_1 \frac{c^2}{2} \, |\mu_1 + \mu_2 - \mu_3 - \mu_4| \, , \tag{6.13}$$

as can be easily verified. In order to assure energy conservation the last quantity must be equal to the differences

$$\mathcal{E} = |\epsilon_1 + \epsilon_2 - \epsilon_3 - \epsilon_4| \tag{6.14}$$

of the chemical link energies ϵ_p of the four different gas species. Equating (6.13) to (6.14) provides the identification of the parameter c

$$c = \sqrt{\frac{2}{m_1} \, \frac{|\epsilon_1 + \epsilon_2 - \epsilon_3 - \epsilon_4|}{|\mu_1 + \mu_2 - \mu_3 - \mu_4|}} \, , \tag{6.15}$$

which is fixed once and for all when the four chemical species are chosen.

The thermodynamic equilibrium in presence of reversible chemical reactions means that one has, in addition to classical equilibrium due to mechanical collisions, the same number of reactions in both directions. In other words for bi-molecular reactions one has, globally in space, the same number of encounters

$$(1,2) \longrightarrow (3,4) \quad , \quad (3,4) \longrightarrow (1,2) \, .$$

Under these conditions, the theory [13] of reacting gases predicts that the ratio of the numerical densities products in opposite sides of reaction is equal to a function Θ depending on the gas temperature only (mass-action law)

$$\frac{\nu_1 \nu_2}{\nu_3 \nu_4} = \Theta(T) \ . \tag{6.16}$$

On the other hand, the discrete kinetic theory of inert gases, developed in Chapter 1, states that in equilibrium the right-hand-side of the kinetic equations vanishes. In other words, referring to Eq.(6.12), the terms,

$$G_i^{pp} + G_i^{pr} - L_i^{pp} - L_i^{pr}$$

due to mechanical collisions, must be equal to zero $\forall i = 1,\ldots,6$ and $\forall p,r = 1,\ldots,4$.

This occurs for the Broadwell model for four gas-components if and only if the equilibrium number densities are expressed by

$$\{\widehat{N}_1^p,\ldots,\widehat{N}_6^p\} = A_p\{e^{c_x}, e^{c_y}, e^{c_z}, e^{-c_x}, e^{-c_y}, e^{-c_z}\} \ , \tag{6.17}$$

where the seven Maxwellian parameters c_x, c_y, c_z, A_p, $p = 1,\ldots,4$ are related to the seven independent macroscopic observables of the gas mixture, namely the four mass densities (one per gas-species) and the three momentum components of the overall mixture.

Moreover, as usual, if at least one of the parameters c_x, c_y, c_z is different from zero, expressions (6.17) provide Maxwellians with drift velocity. In the opposite case ($c_x = c_y = c_z = 0$) one has absolute Maxwellians.

In particular, it can be shown, dealing with chemical equilibrium, that in the presence of bi-molecular reactions only absolute Maxwellians satisfy at the same time the mass-action law (6.16) and the thermodynamic equilibrium in presence of elastic interactions only

$$\forall i = 1, \ldots, 6 \ : \quad \mathcal{G}_i^{pr} - \mathcal{L}_i^{pr} = 0 \ . \tag{6.18}$$

In fact substituting (6.17), with c_x, c_y, c_z different from zero, into the expressions $\mathcal{G}_i^{pr}$ and $\mathcal{L}_i^{pr}$, given by (6.9–6.11), after simple calculations, one obtains the following general expressions for $p = 1$ and $r = 2$ or $p = 3$ and $r = 4$

$$\mathcal{G}_i^{pr} - \mathcal{L}_i^{pr} = \delta_{(p+2)(r+2)} A_{p+2} A_{r+2} \mathcal{R}_i(c_x, c_y, c_z) - \\ - \delta_{pr} A_p A_r \mathcal{H}_i(c_x, c_y, c_z) \ , \tag{6.19}$$

where the terms $\mathcal{R}_i$ and $\mathcal{H}_i$ are functions of the quantities c_x, c_y, c_z which take different forms varying the index i. For this reason, a relationship between the products of the Maxwellian parameters $A_1 A_2$ and $A_3 A_4$, which assures the vanishing of all the expressions (6.19), does not exist. This means that no chemical equilibrium can be recovered in presence of Maxwellians with drift velocity. On the other hand, setting $c_x = c_y = c_z = 0$, leads to

$$\forall i = 1, \ldots, 6 : \quad \mathcal{R}_i = \text{const.} \ , \quad \mathcal{H}_i = \text{const.}$$

and the following relations holds

$$\frac{A_1 A_2}{A_3 A_4} = \frac{K_2}{K_1} \frac{S_{34}[\mu_3 + \mu_4 + (\mu_3^2 + \mu_4^2)^{\frac{1}{2}}]}{S_{12}[\mu_1 + \mu_2 + (\mu_1^2 + \mu_2^2)^{\frac{1}{2}}]} \ , \tag{6.20}$$

which assures that all the terms $\mathcal{G}_i^{pr} - \mathcal{L}_i^{pr}$ vanish in the kinetic equations (6.12).

Eq.(6.20) corresponds to the mass-action law stated by Eq.(6.16), because

$$\Theta(T) = \frac{\nu_1 \nu_2}{\nu_3 \nu_4} = \frac{A_1 A_2}{A_3 A_4}$$

and the right-hand-side turns out to be, through $K_1(T)$ and $K_2(T)$, a function $\Theta(T)$ of the temperature only.

In addition, the knowledge of the function $\Theta(T)$ allows the explicit calculation of the heat Q produced or adsorbed by the chemical reaction. In fact, from the Helmoltz law, one has

$$Q = \mathcal{R}T^2 \frac{d}{dT} \log[\Theta(T)] = \mathcal{R}T^2 \frac{K_1(T)}{K_2(T)} \frac{d}{dT}\left(\frac{K_2}{K_1}\right) , \qquad (6.21)$$

where $\mathcal{R}$ is the constant of gases, with $K_1(T)$ and $K_2(T)$ explicitly given by the model (6.1). In particular the sign of Q from Eq.(6.21) fixes the esothermic or endothermic nature of the chemical reaction: if Q turns out to be positive (heat production) then the reaction is esothermic, otherwise it is endothermic (heat absorption).

6.3 A Mathematical Model for a Gas with Chemical Dissociation and Recombination

In this section, our attention will be directed towards chemical reactions with molecule dissociation and recombination of the type

$$A_2 + A_2 \longrightarrow 4A$$

$$A_2 + A_2 \longrightarrow A_2 + 2A$$

$$A_2 + A \longrightarrow 3A$$

$$A + A \longrightarrow A_2 ,$$

which occur, as we have seen in Section 6.1, for all the halogen gases [14].

In order to deduce the mathematical model of a gas with these chemical reactions, a suitable discrete velocity model must be chosen. In fact, it is necessary to consider a velocity discretization capable to assure momentum conservation for all the chemical reactions considered above. In other words, it is necessary to assure momentum conservation also when the products of the chemical reactions are constituted by an odd number of particles.

Consider then a diatomic gas A_2 in presence of its monoatomic particles A. Because of chemical reactions, the molecules A_2 can dissociate into atoms A and, conversely, atoms A can recombine themselves into molecules A_2. According to this phenomenology the collisional scheme is the following

1) Mechanical collisions (without chemical reactions)

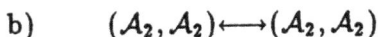

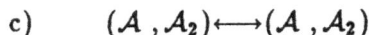

    a) $(A, A) \longleftrightarrow (A, A)$

    b) $(A_2, A_2) \longleftrightarrow (A_2, A_2)$

    c) $(A, A_2) \longleftrightarrow (A, A_2)$

with conservation of momentum and kinetic energy.

2) Collision with chemical dissociation

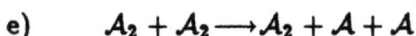

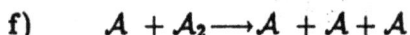

    d) $A_2 + A_2 \longrightarrow A + A + A + A$

    e) $A_2 + A_2 \longrightarrow A_2 + A + A$

    f) $A + A_2 \longrightarrow A + A + A$

with conservation of momentum only.

3) Collisions with chemical recombination

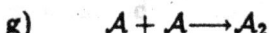

    g) $A + A \longrightarrow A_2$

again with conservation of momentum only.

Energy conservation of chemical interactions will then be assured by a suitable identification of the velocity modulus c, following the same line of Section 6.2.2.

In order to derive the model we will assume that, besides the statement of item (i) of Section 6.2, the following statement holds:

— Collisions (a) and (g), which have the same encounters, provide products with rates $(1 - K_1)$ and K_1, respectively. In the same fashion collisions (b) and (d)–(e) have rates $(1 - K_2)$ and K_2, respectively. Finally collisions (c) and (f) are characterized by rates $(1 - K_3)$ and K_3, respectively. The rates $K_1 \leq 1$, $K_2 \leq 1$, $K_3 \leq 1$ are assumed to be given functions of the local temperature T of the gas. This means that the chemical affinity coefficients K_j, $j = 1, 2, 3$ can be modelled in the same fashion of those of Section 6.2

$$K_j = r_j T^\xi e^{-s_j/T} \quad , \quad j = 1, 2, 3 .$$

A model which allows all the chemical reactions (d)–(g) is the generalization to gas mixtures of the planar 6–velocity model introduced in Section 2.1.3. We will refer then to this model dropping the terms including triple collisions as these ones are not concerned in the above collisional scheme. We remind here that in such a model the selected velocities in the plane are along the directions

$$i = 1, \ldots, 6 : \quad \mathbf{e}_i = \cos(i-1)\frac{\pi}{3}\mathbf{i} + \sin(i-1)\frac{\pi}{3}\mathbf{j}$$

and that the ratio of the moduli has to be properly chosen in order to allow the interactive collisions (c) with momentum conservation. In this case the atoms $\mathcal{A}$ of mass m may attain velocities

$$i = 1, \ldots, 6 : \quad \mathbf{v}_i = c\mathbf{e}_i \ ,$$

while the molecules $\mathcal{A}_2$ of mass $2m$ may attain velocities

$$i = 1, \ldots, 6 : \quad \mathbf{w}_i = \frac{c}{2}\mathbf{e}_i \ .$$

To each set of velocities we join the corresponding number densities (number of particles which in unit time and volume have velocities $\mathbf{v}_i$ and $\mathbf{w}_i$, respectively)

$$N_i = N_i(t, \mathbf{x}) \quad , \quad M_i = M_i(t, \mathbf{x}) \ , \quad \mathbf{x} \in \mathbf{R}^2 \ .$$

It is worth pointing out that in the collisions with chemical reactions (d)–(g), a part of mechanical kinetic energy is transformed into chemical link energy and vice versa. Referring, for instance, to reaction (g) we note that before collision the kinetic energy of the two $\mathcal{A}$–particles is equal to mc^2 while after collision the kinetic energy of the recombined molecule is equal to $\frac{1}{4}mc^2$. Thus we can say that the difference $\frac{3}{4}mc^2$ is the energy $\mathcal{E}$ of the chemical link of the molecule $\mathcal{A}_2$. It is easy to see that the same balance is also verified by reactions (d), (e) and (f). Therefore, in the same fashion of Eq.(6.15), the constant parameter c can be identified by means of the chemical link energy of the considered diatomic gas-molecules

$$c = \sqrt{\frac{4}{3}\frac{\mathcal{E}}{m}} \ .$$

The derivation of the kinetic equations can be realized following a procedure analogous to the one used in Section 6.2.1. Therefore the streaming operators on N_i and M_i are equated to the difference between the loss and gain terms due to both mechanical collisions and chemical reactions. The evolution equations then assume the form

$$\left(\frac{\partial}{\partial t} + \mathbf{v}_i \cdot \nabla_{\mathbf{x}}\right) N_i = J_i^1[\mathbf{N}, \mathbf{M}] + S_i^1[\mathbf{N}, \mathbf{M}] - D_i^1[\mathbf{N}, \mathbf{M}] \qquad (6.22a)$$

$$\left(\frac{\partial}{\partial t} + \mathbf{w}_i \cdot \nabla_{\mathbf{x}}\right) M_i = J_i^2[\mathbf{N}, \mathbf{M}] + S_i^2[\mathbf{N}, \mathbf{M}] - D_i^2[\mathbf{N}, \mathbf{M}] \ , \qquad (6.22b)$$

where

$$\mathbf{N} = \{N_1 \ldots N_6\} \quad , \quad \mathbf{M} = \{M_1 \ldots M_6\}$$

In Eqs.(6.22) the terms J_i^1 and J_i^2 are the collisional operators related to mechanical interactions. The terms S_i^1 and S_i^2 are the *source* operators and D_i^1 and D_i^2 are the *sink* operators due to chemical reactions. A detailed calculation of the above terms will be now performed for each type of collision (a)–(g).

6.3.1 Kinetic Equations of the Discrete Model with Chemical Dissociation and Recombination

The whole set of collisions can be schematized as follows

— *Mechanical Collisions*

The scheme of all admissible mechanical collisions is

a) $(\mathcal{A}, \mathcal{A}) \longleftrightarrow (\mathcal{A}, \mathcal{A})$

The admissible collisions are of the type

$$i = 1, \ldots, 6 \ : \ (N_i, N_{i+3}) \longleftrightarrow \begin{cases} (N_{i+1}, N_{i+4}) \\ \\ (N_{i+2}, N_{i+5}) \end{cases}$$

with relative velocity of the colliding particles equal to $2c$, probability density equal to $\frac{1}{3}$ and rate $(1 - K_1)$.

b) $(\mathcal{A}_2, \mathcal{A}_2) \longleftrightarrow (\mathcal{A}_2, \mathcal{A}_2)$

Similarly to case (a) the admissible collisions are

$$i = 1, \ldots, 6 \ : \ (M_i, M_{i+3}) \longleftrightarrow \begin{cases} (M_{i+1}, M_{i+4}) \\ \\ (M_{i+2}, M_{i+5}) \end{cases}$$

with relative velocity c, probability density $\frac{1}{3}$ and rate $(1 - K_2)$.

c) $(A, A_2) \longleftrightarrow (A, A_2)$

We distinguish between

c$_1$) Head-on collisions

$$i = 1, \ldots, 6 \ : \ (N_i, M_{i+3}) \longleftrightarrow \begin{cases} (N_{i+1}, M_{i+4}) \\[1ex] (N_{i+4}, M_{i+1}) \\[1ex] (N_{i+2}, M_{i+5}) \\[1ex] (N_{i+5}, M_{i+2}) \\[1ex] (N_{i+3}, M_i) \end{cases}$$

with relative velocity $\frac{3}{2}c$, probability density $\frac{1}{6}$ and rate $(1 - K_3)$.

c$_2$) Collisions at angle

$$i = 1, \ldots, 6 \ , \ j \neq i, i + 3 \ : \ (N_i, M_j) \longleftrightarrow (N_j, M_i)$$

with relative velocity

$$R_{ij} = \left[\frac{5}{4} - \cos(i - j)\frac{\pi}{3} \right]^{1/2} c \ ,$$

probability density $\frac{1}{2}$ and rate $(1 - K_3)$.

The expressions of the terms J_i^1 and J_i^2 to be casted into Eqs.(6.22) are then given by

$$J_i^1[\mathbf{N},\mathbf{M}] = \frac{2}{3}(1 - K_1)cS_{11}(N_{i+1}N_{i+4} + N_{i+2}N_{i+5} - 2N_iN_{i+3})$$

$$+ \frac{1}{4}(1 - K_3)cS_{12}\sum_{j=1}^{5}(N_{i+j}M_{i+j+3} - N_iM_{i+3})$$

$$+ \frac{1}{2}(1 - K_3)S_{12}\sum_{\substack{j=1 \\ j\neq i,i+3}}^{6} R_{ij}(M_iN_j - N_iM_j)$$

$$\text{(6.23a)}$$

$$J_i^2[\mathbf{N},\mathbf{M}] = \frac{1}{3}(1 - K_2)cS_{22}(M_{i+1}M_{i+4} + M_{i+2}M_{i+5} - 2M_iM_{i+3})$$

$$+ \frac{1}{4}(1 - K_3)cS_{12}\sum_{j=1}^{5}(M_{i+j}N_{i+j+3} - M_iN_{i+3})$$

$$+ \frac{1}{2}(1 - K_3)S_{12}\sum_{\substack{j=1 \\ j\neq i,i+3}}^{6} R_{ij}(N_iM_j - M_iN_j) \ ,$$

$$\text{(6.23b)}$$

where S_{11}, S_{22} and S_{12} denote respectively the collisional cross-sectional areas in the interaction between atoms, molecules and atoms/molecules.

— *Collisions with Chemical Reactions*

The computation of the collisional terms due to chemical reactions can be schematized as follows

d) $A_2 + A_2 \longrightarrow A + A + A + A$.

We distinguish between

d_1) Head-on collisions

$$i = 1, \ldots, 6 \ : \ (M_i, M_{i+3}) \longrightarrow \begin{cases} (N_i, N_i, N_{i+3}, N_{i+3}) \\[2mm] (N_i, N_{i+1}, N_{i+3}, N_{i+4}) \\[2mm] (N_i, N_{i+2}, N_{i+3}, N_{i+5}) \\[2mm] (N_{i+1}, N_{i+1}, N_{i+4}, N_{i+4}) \\[2mm] (N_{i+1}, N_{i+2}, N_{i+4}, N_{i+5}) \\[2mm] (N_{i+2}, N_{i+2}, N_{i+5}, N_{i+5}) \end{cases}$$

with relative velocity c and rate K_2.

d_2) Collisions at angle

$$i = 1, \ldots, 6 \, , \, j \neq i, i+3 : \ (M_i, M_j) \longrightarrow \begin{cases} (N_i, N_i, N_{i+3}, N_j) \\[2mm] (N_i, N_{i+1}, N_{i+4}, N_j) \\[2mm] (N_i, N_{i+2}, N_{i+5}, N_j) \end{cases}$$

with relative velocity $\frac{1}{\sqrt{2}} R_{ij}$ and rate K_2.

e) $\qquad A_2 + A_2 \longrightarrow A_2 + A + A$.

These collisions are of the type

$$i = 1, \ldots, 6 \ , \ \ j = 1, \ldots, 3 \ : \ (M_i, M_{i+2}) \longrightarrow \begin{cases} (M_{i+1}, N_j, N_{j+3}) \\[2mm] (N_{i+1}, M_j, N_{j+3}) \end{cases}$$

with relative velocity $\frac{\sqrt{3}}{2}c$ and rate K_2.

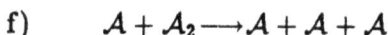

f) $A + A_2 \longrightarrow A + A + A$

Such collisions are

$$i = 1, \ldots, 6 , \quad j \neq i : \quad (N_i, M_j) \longrightarrow (N_{i-1}, N_{i+1}, N_j)$$

with relative velocity R_{ij} and rate K_3.

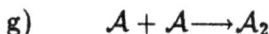

g) $A + A \longrightarrow A_2$

These collisions are of the type

$$i = 1, \ldots, 6 : \quad \begin{cases} (N_i, N_{i+2}) \longrightarrow M_{i+1} \\[2mm] (N_i, N_{i+4}) \longrightarrow M_{i+5} \end{cases}$$

with relative velocity $\sqrt{3}c$ and rate K_1.

Note that the encounters of type (M_i, M_{i+2}) occur both in collisions (d_2) (for $j = i + 2$) and (e). Consequently each encounter will have probability $1/2$ so that the outcomes (d_2) and (e) are equally partitioned.

After this detailed analysis we can calculate each source term S_i and each sink term D_i.

— *Sink terms*

Head-on collisions (d_1) produce sink terms $D_i^{2,1}$ to be inserted in the equations related to the diatomic particles with number densities M_i, namely in Eq.(6.22b). The terms $D_i^{2,1}$ can be expressed by

$$i = 1, \ldots, 6 : \quad D_i^{2,1} = K_2 c S_{22} M_i M_{i+3} . \tag{6.24}$$

Collisions at angle (d$_2$) provide sink terms of molecules $D_i^{2,2}$ in Eq.(6.22b), namely

$$i = 1, \ldots, 6 \ : \quad D_i^{2,2} = \frac{1}{\sqrt{2}} K_2 S_{22} [R_{i,i+1} M_i M_{i+1} + R_{i,i+4} M_i M_{i+4}$$

$$+ R_{i,i+5} M_i M_{i+5} + \frac{1}{2} R_{i,i+2} M_i M_{i+2}] \ .$$

$$(6.25)$$

Collisions (e) provide sink terms of molecules $D_i^{2,3}$ in Eq.(6.22b), namely

$$i = 1, \ldots, 6 \ : \quad D_i^{2,3} = \frac{\sqrt{3}}{4} K_2 c S_{22} M_i M_{i+2} \ . \qquad (6.26)$$

Interactive collisions (f) provide sink terms of molecules $D_i^{2,4}$ and sink terms of atoms $D_i^{1,4}$

$$i = 1, \ldots, 6 \ : \quad D_i^{2,4} = K_3 S_{12} \sum_{\substack{j=1 \\ j \neq i}}^{6} R_{ij} M_i N_j \qquad (6.27)$$

$$i = 1, \ldots, 6 \ : \quad D_i^{1,4} = K_3 S_{12} \sum_{\substack{j=1 \\ j \neq i}}^{6} R_{ij} N_i M_j \qquad (6.28)$$

to be inserted in Eq.(6.22b) and Eq.(6.22a), respectively.

Collisions at angle (g) provide sink terms of atoms $D_i^{1,5}$ to be inserted in Eq.(6.22a), given by

$$i = 1, \ldots, 6 \ : \quad D_i^{1,5} = \sqrt{3} K_1 c S_{11} N_i (N_{i+2} + N_{i+4}) \ . \qquad (6.29)$$

Accordingly in Eq.(6.22a) the sink terms are given by

$$D_i^1 = D_i^{1,4} + D_i^{1,5} \qquad (6.30)$$

and in Eq.(6.22b) by

$$D_i^2 = D_i^{2,1} + D_i^{2,2} + D_i^{2,3} + D_i^{2,4} \ . \qquad (6.31)$$

— *Source terms*

In order to assure total mass conservation it is necessary that the sum of all the sinks contributions must be equal to the sum of all the source terms. Thus collisional products arising from encounters (d)–(g) must be introduced as source terms in the equations of the species produced, that is atoms in Eq.(6.22a), molecules in Eq.(6.22b). According to the outcomes defined by the collisional scheme (d)–(g), such products of atoms and molecules will then be equally distributed in the appropriate equations as source terms.

Thus $\sum_i D_i^{2,1}$ will be equally distributed in the set of Eq.(6.22a), giving

$$i = 1, \ldots, 6 : \quad S_i^{1,1} = \frac{1}{3} K_2 c S_{22} \sum_{k=1}^{3} M_k M_{k+3} \ . \qquad (6.32)$$

In the same fashion $\sum_i D_i^{2,2}$ will be distributed again in Eq.(6.22a), so that

$$i = 1, \ldots, 6 : \quad S_i^{1,2} = \frac{K_2 S_{22}}{6\sqrt{2}} \sum_{k=1}^{6} [R_{k,k+1} M_k M_{k+1} + R_{k,k+4} M_k M_{k+4}$$

$$+ R_{k,k+5} M_k M_{k+5} + \frac{1}{2} R_{k,k+2} M_k M_{k+2}] .$$

(6.33)

Moreover $\sum_i D_i^{2,3}$ must be distributed in both Eqs.(6.22) in different rates, since encounters (M_i, M_{i+2}) give rise to products

$$(M_{i+1}, N_j, N_{j+3}) \quad \text{and} \quad (N_{i+1}, M_j, N_{j+3}) \ , \quad j = 1, 2, 3 .$$

Thus 1/3 of the above sum will be counted in Eq.(6.22b) while 2/3 in Eq.(6.22a). Then

$$i = 1, \ldots, 6 : \quad S_i^{1,3} = \frac{\sqrt{3}}{36} K_2 c S_{22} \sum_{k=1}^{6} M_k M_{k+2}, \qquad (6.34)$$

$$i = 1, \ldots, 6 : \quad S_i^{2,3} = \frac{\sqrt{3}}{72} K_2 c S_{22} \sum_{k=1}^{6} M_k M_{k+2}. \qquad (6.35)$$

Again both contributions $\sum_i D_i^{1,4}$ and $\sum_i D_i^{2,4}$ will be partitioned in Eq.(6.22a), so that

$$i = 1, \ldots, 6 : \quad S_i^{1,4} = \frac{1}{3} K_3 S_{12} \sum_{k=1}^{6} \sum_{\substack{j=1 \\ j \neq k}}^{6} R_{kj} M_k N_j . \qquad (6.36)$$

Finally $\sum_i D_i^{1,5}$ will be distributed in Eq.(6.22b), giving

$$i = 1, \ldots, 6 : \quad S_i^{2,5} = \frac{\sqrt{3}}{6} K_1 c S_{11} \sum_{k=1}^{6} N_k (N_{k+2} + N_{k+4}) . \qquad (6.37)$$

In conclusion the source terms in Eq.(6.22a) are given by

$$S_i^1 = S_i^{1,1} + S_i^{1,2} + S_i^{1,3} + S_i^{1,4} \tag{6.38}$$

while in Eq.(6.22b) the source terms are

$$S_i^2 = S_i^{2,3} + S_i^{2,5} . \tag{6.39}$$

The kinetic equations can be now written in a final form by inserting in Eqs.(6.22) the mechanical collisional terms (6.23) and the contributions due to chemical reactions (6.24–6.39). Thus

$$\frac{\partial N_i}{\partial t} + c e_i \cdot \nabla_{\mathbf{x}} N_i$$

$$= \frac{2}{3}(1 - K_1)cS_{11}(N_{i+1}N_{i+4} + N_{i+2}N_{i+5} - 2N_iN_{i+3})$$

$$+ \frac{1}{4}(1 - K_3)cS_{12} \sum_{j=1}^{5}(N_{i+j}M_{i+j+3} - N_iM_{i+3})$$

$$+ \frac{1}{2}(1 - K_3)S_{12} \sum_{\substack{j=1 \\ j \neq i, i+3}}^{6} R_{ij}(M_iN_j - N_iM_j)$$

$$+ \frac{1}{3}K_2cS_{22} \sum_{j=1}^{3} M_jM_{j+3} + \frac{1}{6\sqrt{2}}K_2S_{22}$$

$$\cdot \sum_{j=1}^{6}(R_{j,j+1}M_{j+1} + R_{j,j+4}M_{j+4} + R_{j,j+5}M_{j+5} + \frac{1}{2}R_{j,j+2}M_{j+2})M_j$$

$$\tag{6.40a}$$

$$+ \frac{\sqrt{3}}{36} K_2 c S_{22} \sum_{j=1}^{6} M_j M_{j+2} + \frac{1}{3} K_3 S_{12} \sum_{k=1}^{6} \sum_{\substack{k=1 \\ j \neq k}}^{6} R_{kj} M_k N_j$$

$$- K_3 S_{12} \sum_{\substack{j=1 \\ j \neq i}}^{6} R_{ij} N_i M_j - \sqrt{3} K_1 c S_{11} N_i (N_{i+2} + N_{i+4})$$

$$\frac{\partial M_i}{\partial t} + \frac{1}{2} c e_i \cdot \nabla_x M_i$$

$$= \frac{1}{3} (1 - K_2) c S_{22} (M_{i+1} M_{i+4} + M_{i+2} M_{i+5} - 2 M_i M_{i+3})$$

$$+ \frac{1}{4} (1 - K_3) c S_{12} \sum_{j=1}^{5} (M_{i+j} N_{i+j+3} - M_i N_{i+3})$$

$$+ \frac{1}{2} (1 - K_3) S_{12} \sum_{\substack{j=1 \\ j \neq i, i+3}}^{6} R_{ij} (N_i M_j - M_i N_j)$$

$$+ \frac{\sqrt{3}}{72} K_2 c S_{22} \sum_{k=1}^{6} M_k M_{k+2} + \frac{\sqrt{3}}{6} K_1 c S_{11} \sum_{k=1}^{6} N_k (N_{k+2} + N_{k+4})$$

$$- \frac{K_2}{\sqrt{2}} S_{22} (R_{i,i+1} M_{i+1} + R_{i,i+4} M_{i+4} + R_{i,i+5} M_{i+5} + \frac{R_{i,i+2}}{2} M_{i+2}) M_i$$

$$- K_2 c S_{22} M_i M_{i+3} - \frac{\sqrt{3}}{4} K_2 c S_{22} M_i M_{i+2} - K_3 S_{12} \sum_{\substack{j=1 \\ j \neq i}}^{6} R_{ij} M_i N_j \ .$$

$$(6.40b)$$

6.3.2 Thermodynamic Equilibrium in Presence of Chemical Dissociation and Recombination

Assume that the gas is in absolute Maxwellian equilibrium

$$i = 1, \ldots, 6 \; : \; \widehat{N}_i = A_1 \;\; , \;\; \widehat{M}_i = A_2 \;\; , \;\; A_1, A_2 \in \mathbf{R}_+ \; ,$$

so that the pure mechanical collision operators J_i^1 and J_i^2 in (6.23) vanish.

On the other hand thermodynamic equilibrium in the presence of chemical reactions implies that in the whole gas the number of dissociating particles equals the number of recombination reactions. Therefore, in order to obtain such a chemical equilibrium, it is necessary to find the relation between A_1 and A_2.

For this purpose casting A_1 and A_2 in the kinetic equations (6.40) leads to

$$\begin{aligned}
S_i^{(1)} - D_i^{(1)} = -(S_i^{(2)} - D_i^{(2)}) = \\
= -\mathcal{H}_1(T)A_1^2 + \mathcal{H}_2(T)A_1 A_2 + \mathcal{H}_3(T)A_2^2
\end{aligned} \tag{6.41}$$

$\forall i = 1, \ldots, 6$, where

$$\mathcal{H}_1(T) = 2\sqrt{3}K_1(T)cS_{11}$$

$$\mathcal{H}_2(T) = \frac{3 + 2\sqrt{3} + 2\sqrt{7}}{2}K_3(T)cS_{12}$$

$$\mathcal{H}_3(T) = \left[1 + \frac{3}{4}\sqrt{\frac{7}{2}} + \left(\frac{2}{\sqrt{2}} + \frac{1}{6} \right)\sqrt{3} \right] K_2(T)cS_{22}$$

are known functions of temperature T.

Thus, setting Eq.(6.41) equal to zero, gives the equilibrium rate

$$\Theta(T) = \frac{A_2}{A_1} = \frac{(\mathcal{H}_2 + 4\mathcal{H}_1\mathcal{H}_3)^{1/2} - \mathcal{H}_2}{2\mathcal{H}_3} .$$

$\Theta(T)$ represents the rate ν_2/ν_1 between the numerical densities of molecules and atoms in the states of thermodynamic and chemical equilibrium at temperature T. Such a rate will be uniquely determined once the expressions of K_j, $j = 1, \ldots, 3$, as functions of temperature are provided, according to the rules $K_j = r_j T^\xi e^{-s_j/T}$ introduced above.

6.4   A Discrete Model for a Dissociating Gas

This section deals with the proposal of a three-dimensional model of the discrete Boltzmann equation which describes the behaviour of a gas where only chemical reactions of dissociation occur. The aim of this proposal is to apply this model to the problem of shock wave onset, already dealt with in Chapter 4. This application will then be dealt with in Section 6.5.

Consider then a dilute gas mixture of atoms A and diatomic molecules A_2 which can dissociate. The gas particles admit collisions of the following type

1) Mechanical binary collisions (without chemical reactions)
    a) $(A, A) \longleftrightarrow (A, A)$
    b) $(A_2, A_2) \longleftrightarrow (A_2, A_2)$

c) $(\mathcal{A}, \mathcal{A}_2) \longleftrightarrow (\mathcal{A}, \mathcal{A}_2)$

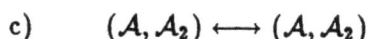

with conservation of momentum and energy.

2) Collisions with chemical dissociation

d) $\mathcal{A}_2 + \mathcal{A}_2 \longrightarrow \mathcal{A} + \mathcal{A} + \mathcal{A} + \mathcal{A}$

with momentum conservation only.

Collisions (e), (f) and (g) of Section 6.3 are not considered here.

Moreover, referring to the rates of dissociation and to the internal degrees of freedom, the same hypothesis introduced in Section 6.3 will be assumed. Therefore referring to collisions of the type $(\mathcal{A}_2, \mathcal{A}_2)$ a particle fraction with rate $K \leq 1$ interacts according to the chemical reaction (d), while the other fraction $(1 - K)$ interacts according to the merely mechanical scattering (b). The rate K is modelled as a known function of the gas temperature, $K = K(T)$.

In order to obtain a specific model describing the behaviour of the dissociating gas, one needs to define a velocity discretization so that the mechanics of collisions and of chemical reactions can be described, keeping in mind the assumptions (i)–(iv) of Section 6.2. Namely the discretization of the velocity directions is the one defined by the six unit vectors e_i, $i = 1, \ldots, 6$, defined by an orthogonal frame in the space with the directions of the three axes and their opposite ones.

In particular, with obvious meaning of symbols we have that the selected velocities of the particles are

$v_i = ce_i$ for the atoms $\mathcal{A}$ of mass m and number densities N_i

$w_i = \frac{1}{2}ce_i$ for the molecules $\mathcal{A}_2$ of mass $2m$ and number densities M_i.

Referring to the velocity discretization, two non-independent velocity moduli c and $\frac{1}{2}c$ imply that collisions (a)–(c) satisfy energy and momentum conservation. Moreover in the chemical reaction (d) the kinetic energy before collision is $\frac{1}{2}mc^2$, while after collision is $2mc^2$. Therefore the mechanical energy gain is equal to $\frac{3}{2}mc^2$ and must be interpreted as the chemical link energy lost by the two molecules A_2. Thus c is uniquely identified by $\sqrt{\frac{4\mathcal{E}}{3m}}$, $\mathcal{E}$ being the chemical link energy of the gas molecule.

The kinetic equations can be as usual derived equating the streaming operator of N_i and M_i to the collisional operator due to mechanical interactions as well as to chemical reactions. Thus using the same formalism of Eqs.(6.22) the evolution equations are given by

$$\frac{\partial N_i}{\partial t} + \mathbf{v}_i \cdot \nabla_{\mathbf{x}} N_i = J_i^1[\mathbf{N}, \mathbf{M}] + S_i[\mathbf{M}]$$

$$\frac{\partial M_i}{\partial t} + \mathbf{w}_i \cdot \nabla_{\mathbf{x}} M_i = J_i^2[\mathbf{N}, \mathbf{M}] - D_i[\mathbf{M}] , \qquad i = 1, \dots, 6 \qquad (6.42)$$

where

$$\mathbf{N} = \{N_1, \dots, N_6\} \ , \quad \mathbf{M} = \{M_1, \dots, M_6\} \ ,$$

J_i^1 and J_i^2 are the mechanical collision operators and S_i and D_i are, respectively, the source and sink terms due to chemical dissociation.

The expressions of J_i^1 and J_i^2 are the ones of the Broadwell model for a binary gas mixture, already dealt with in Chapter 2, and here rewritten considering that only the rate $(1 - K)$ of the collision between molecules have a mechanical interaction

$$
J_i^1[\mathbf{N}, \mathbf{M}] = \frac{2}{3} c S_{11} (N_{i+1} N_{i+4} + N_{i+2} N_{i+5} - 2 N_i N_{i+3})
$$
$$
+ \frac{c S_{12}}{4} \{ N_{i+1} M_{i+4} + N_{i+4} M_{i+1} + N_{i+2} M_{i+5}
$$
$$
+ N_{i+5} M_{i+2} + N_{i+3} M_i - 5 N_i M_{i+3} \qquad (6.43a)
$$
$$
+ \sqrt{5} [M_i (N_{i+1} + N_{i+2} + N_{i+4} + N_{i+5})
$$
$$
- N_i (M_{i+1} + M_{i+2} + M_{i+4} + M_{i+5})] \}
$$

$$
J_i^2[\mathbf{N}, \mathbf{M}] = \frac{1}{3} c S_{22} (1 - K) (M_{i+1} M_{i+4} + M_{i+2} M_{i+5} - 2 M_i M_{i+3})
$$
$$
+ \frac{c S_{12}}{4} \{ M_{i+1} N_{i+4} + M_{i+4} N_{i+1} + M_{i+2} N_{i+5}
$$
$$
+ M_{i+5} N_{i+2} + M_{i+3} N_i - 5 M_i N_{i+3}
$$
$$
+ \sqrt{5} [N_i (M_{i+1} + M_{i+2} + M_{i+4} + M_{i+5})
$$
$$
- M_i (N_{i+1} + N_{i+2} + N_{i+4} + N_{i+5})] \}
$$
$$
(6.43b)
$$

where S_{11}, S_{22}, S_{12} are the collisional cross sectional areas of atoms, molecules and atoms-molecules interactions, respectively.

These equations have been reported in the book several times and are repeated in this chapter for sake of completeness and in view of the application dealt with in Section 6.5.

We have now to compute the sources S_i and sinks D_i. The terms S_i correspond to the atoms rising from dissociation. Conversely D_i are terms due to molecules lost in dissociation.

The admissible encounters and collision products which preserve mass and momentum in the chemical interaction (d) are

$$i = 1, \ldots, 6 \quad : \quad (M_i, M_{i+3}) \longleftrightarrow \begin{cases} (N_i, N_{i+3}, N_i, N_{i+3}) \\[2mm] (N_i, N_{i+3}, N_{i+1}, N_{i+4}) \\[2mm] (N_i, N_{i+3}, N_{i+2}, N_{i+5}) \\[2mm] (N_{i+1}, N_{i+4}, N_{i+1}, N_{i+4}) \\[2mm] (N_{i+1}, N_{i+4}, N_{i+2}, N_{i+5}) \\[2mm] (N_{i+2}, N_{i+5}, N_{i+2}, N_{i+5}) \end{cases} \qquad (6.44)$$

with incoming relative velocity equal to c and probability rate K.

Thus, following the standard modelling technique, which leads to the discrete Boltzmann equation and which has been dealt with several times in this book, the terms D_i, to be substituted in Eqs.(6.42), can be written as

$$D_i[\mathbf{M}] = K c S_{22} M_i M_{i+3} . \qquad (6.45)$$

On the other hand, conservation of total mass prescribes that to each sink term a corresponding source term must appear in each equation (6.42). At the same time, the collisional scheme (6.44) implies that the number densities of the outcoming velocities are equally partitioned in every space direction. Therefore each equation must contain as source term an equal fraction of the total sink contribution, that is

$$S_i[\mathbf{M}] = \frac{1}{6} K c S_{22} \sum_{k=1}^{6} M_k M_{k+3} . \qquad (6.46)$$

The model is now completely determined if one substitutes the expressions of the collision operators, defined in Eqs.(6.43, 6.45, 6.46) into the kinetic evolution equations (6.42). The whole set of equations will be reported in the next section for the one dimensional case.

Additional analysis of some thermodynamical properties of this model will be dealt with in the next section.

6.5   Shock Waves for a Dissociating Gas

As an application of the mathematical models described in this chapter, we study here the classical problem of the shock wave formation, widely dealt with in Chapter 4, for the dissociation model proposed in the previous section.

Having in mind the content of Chapter 4, consider then an infinite tube, divided by a diaphragm located at $x = 0$ into two regions, filled by the mixture $A - A_2$ at two different thermodynamical states. At $t = 0$ the diaphragm is instantaneously withdrawn. After a transient time a shock wave front is formed and propagates with steady velocity.

In order to study this well-known problem [14] it is convenient to reduce the equations (6.42–6.46) characterizing the model to the case corresponding to one space dimension

$$N_2 = N_3 = N_5 = N_6 \quad , \quad M_2 = M_3 = M_5 = M_6 \; . \qquad (6.47)$$

Consider then a scaling of time, of space and of the number densities similar to the one used in Chapter 4.

$$\tau = \frac{ct}{\ell} \ , \ \eta = \frac{x}{\ell} \ ,$$

$$f_+ = \frac{N_1}{N} \ , \ f_o = \frac{N_2}{N} \ , \ f_- = \frac{N_4}{N} \ ,$$

$$g_+ = \frac{M_1}{N} \ , \ g_o = \frac{M_2}{N} \ , \ g_- = \frac{M_4}{N} \ ,$$

where $N = \sum\limits_{i=1}^{6} N_i(x \to -\infty)$ and ℓ is a fixed laboratory length scale. Moreover one can define the following quantities

$$Kn_1 = \frac{1}{S_{11}N\ell} \ , \ Kn_2 = \frac{1}{S_{22}N\ell} \ , \ Kn_3 = \frac{1}{S_{12}N\ell} \tag{6.48}$$

which are proportional to three typical Knudsen numbers. Hence, the kinetic equations are reduced to a set of six nonlinear partial differential equations, three of them refer to atoms

$$\frac{\partial f_+}{\partial \tau} + \frac{\partial f_+}{\partial \eta} = \frac{4}{3Kn_1}(f_o^2 - f_+f_-) + \frac{1}{4Kn_3}(4f_og_o + f_-g_+ - 5f_+g_-)$$

$$+ \frac{\sqrt{5}}{Kn_3}(f_og_+ - f_+g_o) + \frac{K}{3Kn_2}(g_+g_- + 2g_o^2)$$

$$\frac{\partial f_o}{\partial \tau} = -\frac{2}{3Kn_1}(f_o^2 - f_+f_-) + \frac{1}{4Kn_3}(f_-g_+ + f_+g_- - 2f_og_o)$$

$$+ \frac{\sqrt{5}}{4Kn_3}(f_-g_o + f_+g_o - f_og_- - f_og_+) + \frac{1}{3Kn_2}K(g_+g_- + 2g_o^2)$$

$$\tag{6.49a}$$

$$\frac{\partial f_-}{\partial \tau} - \frac{\partial f_-}{\partial \eta} = \frac{4}{3Kn_1}(f_o^2 - f_+ f_-) + \frac{1}{4Kn_3}(4f_o g_o + f_+ g_- - 5f_- g_+)$$

$$+ \frac{\sqrt{5}}{Kn_3}(f_o g_- - f_- g_o) + \frac{K}{3Kn_2}(g_+ g_- + 2g_o^2)$$

and three to molecules

$$\frac{\partial g_+}{\partial \tau} + \frac{1}{2}\frac{\partial g_+}{\partial \eta} = \frac{2}{3Kn_2}(1 - K)(g_o^2 - g_+ g_-)$$

$$+ \frac{1}{4Kn_3}(4f_o g_o + f_+ g_- - 5f_- g_+)$$

$$+ \frac{\sqrt{5}}{Kn_3}(f_+ g_o - f_o g_+) - \frac{K}{Kn_2}g_+ g_-$$

$$\frac{\partial g_o}{\partial \tau} = \frac{1}{3Kn_2}(K - 1)(g_o^2 - g_+ g_-)$$

$$+ \frac{1}{4Kn_3}(f_+ g_- + f_- g_+ - 2f_o g_o) \qquad (6.49b)$$

$$+ \frac{\sqrt{5}}{4Kn_3}(f_o g_- + f_o g_+ - f_- g_o - f_+ g_o) - \frac{K}{Kn_2}g_o^2$$

$$\frac{\partial g_-}{\partial \tau} - \frac{1}{2}\frac{\partial g_-}{\partial \eta} = \frac{2}{3Kn_2}(1 - K)(g_o^2 - g_+ g_-)$$

$$+ \frac{1}{4Kn_3}(4f_o g_o + f_- g_+ - 5f_+ g_-)$$

$$+ \frac{\sqrt{5}}{Kn_3}(f_- g_o - f_o g_-) - \frac{K}{Kn_2}g_+ g_- \ .$$

Having in mind the mathematical simulation of the shock-onset problem, consider the initial value problem for Eqs.(6.49) with the following initial conditions

$$\eta \to -\infty \; : \; f_+ = 1 \; , \quad f_- = f_o = 0 \; ,$$
$$g_+ = B \; , \quad g_- = g_o = 0 \; ; \tag{6.50}$$

$$\eta \to +\infty \; : \; f_+ = f_- = f_o = A_1 \; ,$$
$$g_+ = g_- = g_o = A_2 \; . \tag{6.51}$$

Conditions (6.50) correspond to a Maxwellian state with gas temperature equal to zero (infinite Mach number), whereas conditions (6.51) represent a Maxwellian state with zero mean velocity at the temperature

$$\widehat{T}_M = \frac{mc^2}{3k_B} \frac{\nu_1 + \frac{\nu_2}{2}}{\nu_1 + \nu_2} \; ,$$

where $\nu_1 = 6A_1$ and $\nu_2 = 6A_2$. Moreover, we assume the rate K defined by

$$K(T; \widehat{T}_M) = \sqrt{\frac{T}{\widehat{T}_M(2 - \sqrt{3})}} \exp\left[-\frac{\widehat{T}_M}{\widehat{T}_M - T} + \frac{1}{\sqrt{3} - 1} \right] \leq 1 \; , \tag{6.52}$$

according to the fact that the chemical dissociation must be equal to zero in the above equilibrium states.

Such a form has been obtained extending to discrete models a known expression of the rate K proposed by Bird [1]. In other words,

Eq.(6.52) states, at the level of mathematical model, that the dissociation rate is a known function of the local temperature.

As already shown in Chapter 4, the constants A_1 and A_2 in (6.51) can be determined by the Rankine-Hugoniot conditions, together with the constant propagation speed β, as functions of B. In fact, one finds

$$\beta = \frac{2 + B - [(2 + B)^2 + 6(1 + B)(\frac{B}{2} + 2)]^{1/2}}{6(1 + B)} < 0$$

$$A_1 = \frac{\beta - 1}{6\beta} \qquad (6.53)$$

$$A_2 = \frac{\beta - \frac{1}{2}}{6\beta} B \ .$$

Moreover it is possible to express the speed β as a function of the sound speed $c\lambda$ of the binary mixture $\mathcal{A} - \mathcal{A}_2$ at $+\infty$. In fact, from Section 4.2, it turns out that λ is given by

$$\lambda = \sqrt{\frac{\nu_1 + \frac{\nu_2}{4}}{3(\nu_1 + \frac{\nu_2}{2})}} \qquad (6.54)$$

and consequently, after some algebraic manipulations, one obtains

$$\beta = -\ \frac{1 + \frac{B}{4} + 9\lambda^2(1 + \frac{B}{2})}{18\lambda^2(1 + B)}$$
$$-\ \frac{\{[1 + \frac{B}{4} + 9\lambda^2(1 + \frac{B}{2})]^2 + 36\lambda^2(1 + \frac{B}{8})(1 + B)\}^{1/2}}{18\lambda^2(1 + B)}\ .$$

We have now all elements to obtain quantitative results derived from the numerical integration of the system of partial differential equations where the initial data are assumed extending conditions (6.50) to every $\eta \leq 0$ and conditions (6.51) to every $\eta > 0$. In particular if one imposes $B = 2$ then atoms and molecules have $\forall \eta \leq 0$ the same mean velocity $u_1 = u_2$. The numerical scheme which has been adopted is a 2nd-order finite difference code. The reader can verify the calculations by using either a similar computation code or any other equivalent numerical method.

The results of some numerical computations are shown in Figs. 6.1 and 6.2. In details Fig. 6.1 shows the asymptotic (for large values of τ) numerical density profiles of atoms and molecules in presence of dissociation compared with the analogous profiles obtained in absence of chemical reactions. This comparison shows an atom increase, due to molecule dissociation, and a corresponding molecule rarefaction behind the shock-front.

This situation is furtherly confirmed by Fig. 6.2 where the total entropies defined in (1.38) are shown in presence (a) and absence (b) of dissociation, respectively. In fact, the entropy profile in presence of dissociation is characterized by a hump behind the overshoot on the shock-front. Such an entropy rise can be regarded as a measure of a larger disequilibrium state compared with the one of a gas without chemical reactions.

The reader can start from these calculation for a further insight into the description provided by this model of the physical situations related to shock wave propagation of reacting gases.

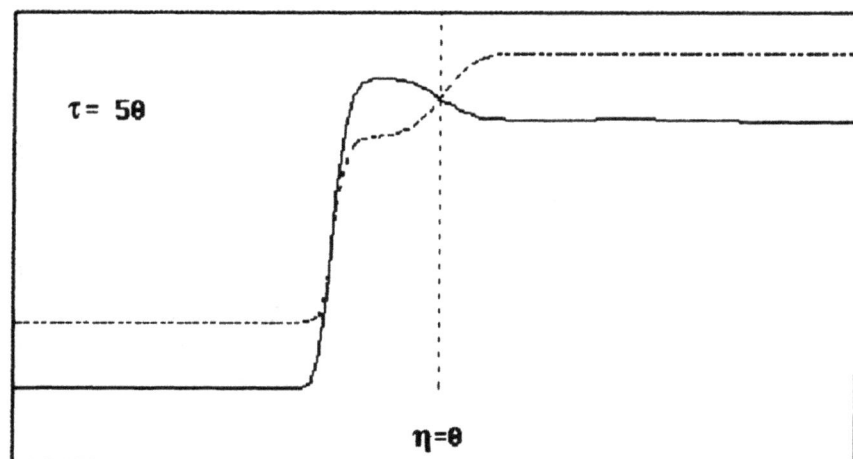

(a)

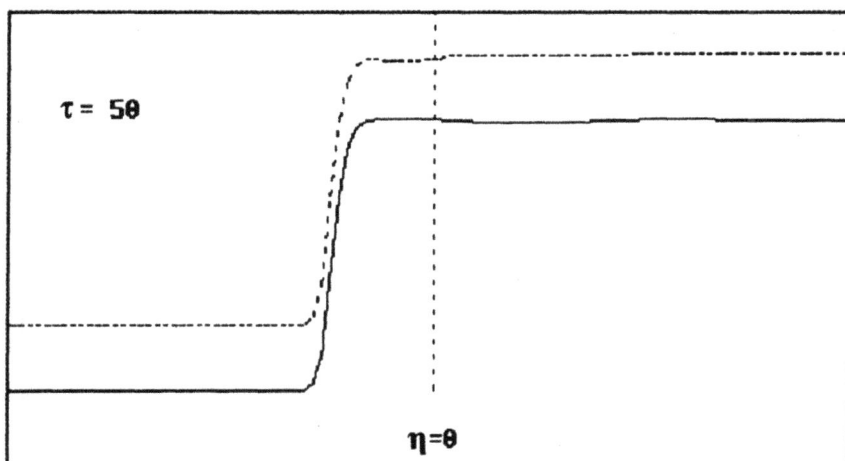

(b)

Fig. 6.1 — Numerical density profiles of atoms (full lines) and molecules (dashed lines), at $\tau = 50$, with (a) and without (b) molecule dissociation. Shock propagation speed $\beta = -.2426$.

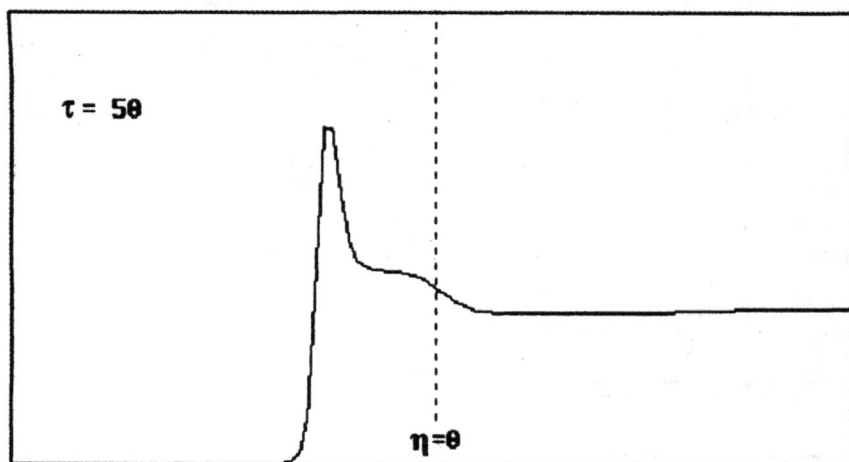

(a)

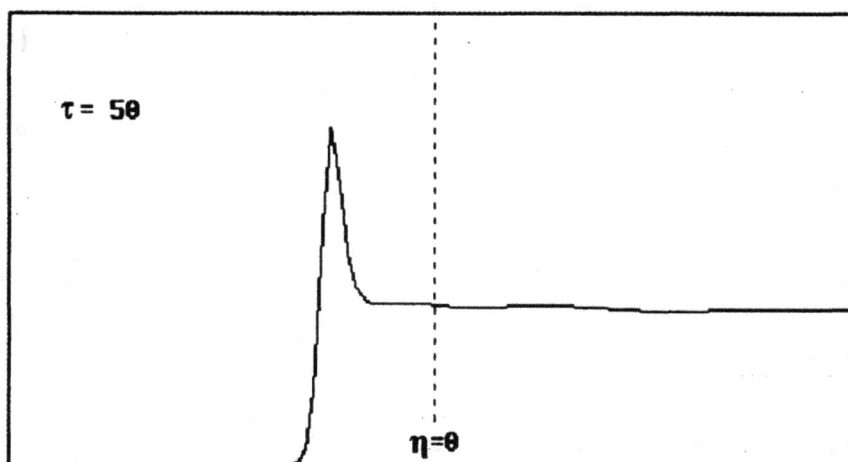

(b)

Fig. 6.2 — *Total entropy profiles at $\tau = 50$ with (a) and without (b) dissociation. Shock propagation speed $\beta = -.2426$.*

6.6   Discussion and Open Problems

The content of this chapter refers, as we have seen, to a topic which only recently has been introduced into the pertinent literature. This topic is certainly interesting and deserves future research activity. As a matter of fact, only three mathematical models of the discrete Boltzmann equation with chemical reactions have been proposed in this chapter and a simple application has been dealt with.

A large amount of research perspectives are then left to future activity of fluid-dynamicists and applied mathematicians.

As we have seen throughout this chapter, the mathematical modelling of the discrete Boltzmann equation with chemical reactions has generated some interesting (but still rather simple) models. Several problems are, at least at present, still open. In fact a general modelling methodology is not yet available as for the classical discrete Boltzmann equation and certainly research in this direction would be useful.

In particular the modelling should consider velocity discretizations more general than the ones dealt with in this chapter with detailed attention to modelling with triple collisions. In fact triple collisions may hopefully be able to simulate catalytic reactions which cannot be simulated in the framework of the classical Boltzmann equation. This difficulty is related, as known [15], to the difficulty in providing explicit models of the Boltzmann equation in the case of multiple collisions.

Nevertheless it is worth mentioning that a model of the continuous Boltzmann equation including triple collisions [16] has been recently proposed with the aim of studying dissociation reactions.

Some analysis of the thermodynamic equilibrium properties have been developed for the models described in this chapter. This must be regarded as a preliminary approach. The whole topic deserves a systematic approach, for general classes of models, as the ones developed in Chapters 1 and 2.

The analysis of such a problem should be completed with the related investigation on the stability properties of the equilibrium states, when they exist.

Moreover several mathematical aspects should still be investigated. In particular we refer to the content of the survey of Chapter 3, i.e. the qualitative analysis of mathematical problems such as (amongst others) the initial value problem in the whole space and the initial-boundary value problem in bounded domains.

As far as the initial value problem is concerned, several important results are available in the literature, as reviewed in Chapter 3, for the classical discrete Boltzmann equation. On the other hand, nothing is known for the discrete Boltzmann equation with chemical reactions.

The analysis of this type of problems cannot be regarded as a natural generalization of the corresponding results for the classical equation. This applies, at most, to local existence theorems. In fact if one refers to global existence results two serious difficulties need to be tackled: the lack of conservation of mechanical energy and the lack of explicit Liapunov functionals (in the classical equation, the H-function).

Applied mathematicians who intend to deal with such a problem may need to give a suitable preliminary answer to the aspect of the problem which have been mentioned above.

The analysis of fluid-dynamic applications of the discrete Boltzmann equation for chemically reacting gases is certainly of great interest in the fluid-mechanics of real gases. In fact these models can explain several real gas effects which are consistent with high velocity and high temperature flows.

A sample application has been given in Section 6.5 but the subject needs to be furtherly developed. In particular, for flows in bounded domains one should consider the formulation of the boundary conditions, in the style indicated in Chapter 5, taking into account the chemical reaction effects involved by the collisions with the walls. These effects can be increased for high impact velocities (as for aerospace vehicles) or for other similar physical situations, for instance high temperature of the wall or reacting solids.

Furthermore initial-boundary value problems may be interesting to simulate, among other physical situations, chemical reactions due to the presence of a solid catalizer.

In conclusion, it can be stated that the content of this chapter opens to several research perspectives which may be hopefully dealt with in the future by applied mathematicians.

References

[1] G.A. Bird, **Molecular Gas Dynamics**, Clarendon Press, Oxford, 1976.

[2] N. Peters and C. Kennel, "Introduction a l'ètude de la combustion", in **Combustion Modelling**, *Advances in Mathematics for Applied Sciences* vol.6, Ed. B. Larrouturou, World Scientific, London, Singapore, 1991, p.121.

[3] N. Xystris and J. Dahler, "A reactive collision model for use in kinetic theory", *J. Chem. Phys.*, **68**, 1978, p.345.

[4] N. Xystris and J. Dahler, "Mass and momentum transport in dilute reacting gases", *J. Chem. Phys.*, **68**, 1978, p.354.

[5] R. Kapral, "Kinetic theory of chemical reactions in liquids", in **Advances in Chemical Physics**, Eds. I. Prigogine and S. Rice, vol. **48**, Wiley and Sons, New York, 1981, p.71.

[6] V. Boffi and A. Rossani, "On the Boltzmann system for a mixture

of reacting gases", *ZAMP*, **41**, 1990, p.254.

[7] N. Xystris and J. Dahler, "Enskog theory for chemically reacting fluids", *J. Chem. Phys.*, **68**, 1978, p.374.

[8] N. Xystris and J. Dahler, "Kinetic theory of simple reacting spheres", *J. Chem. Phys.*, **68**, 1978, p.387.

[9] G. Searby, V. Zehnlé and B. Denet, "Lattice gas mixtures of reactive flows", in **Discrete Kinetic Theory, Lattice Gas Dynamics and Foundations of Hydrodynamics**, Ed. R. Monaco, World Scientific, London, Singapore, 1989, p.300.

[10] R. Monaco and M. Pandolfi Bianchi, "Shock wave onset with chemical dissociation by the discrete Boltzmann equation", in **Rarefied Gas Dynamics**, Ed. A. Beylich, VCH-Verlag, Weinheim, New York, 1991, p.862.

[11] E. Gabetta and R. Monaco, "On the modelling of the discrete Boltzmann equation for gases with bi-molecular chemical reactions", in **Discrete Models of Fluid-Dynamics**, *Advances in Mathematics for Applied Sciences* vol.**2**, Ed. A. Alves, World Scientific, London, Singapore, p.22.

[12] R. Monaco and M. Pandolfi Bianchi, "A discrete velocity model with chemical reactions of dissociation and recombination", in **Advances in Kinetic Theory and Continuum Mechanics**, Eds. R. Gatignol and Soubbaramayer, Springer-Verlag, Berlin, New York, 1991, p.169.

[13] G. Lewis and M. Randall, **Thermodynamics**, McGraw-Hill, New York, 1970.

[14] J.N. Bradley, **Shock Waves in Chemistry and Physics**, John

Wiley Inc., New York, 1962.

[15] N. Bellomo, M. Lachowicz, J. Polewczak and G. Toscani, **Mathematical Topics in Nonlinear Kinetic Theory II: The Enskog Equation**, *Advances in Mathematics for Applied Sciences* vol.1, World Scientific, London, Singapore, 1991.

[16] J. A. McLennan, "Boltzmann equation for a dissociating gas", *J. Statist. Phys.*, **57**, 1989, p.887.

AUTHOR INDEX

SUBJECT INDEX

Series on Advances in Mathematics for Applied Sciences

Aims and Scope

This Series reports on new developments in mathematical research relating to methods, qualitative and numerical analysis, mathematical modeling in the applied and the technological sciences. Contributions rlated to constitutive theories, fluid dynamics, kinetic and transport theories, solid mechanics, system theory and mathematical methods for the applications are welcomed.

This Series includes books, lecture notes, proceedings, collections of research papers. Monograph collections on specialized topics of current interest are particularly encouraged. Both the proceedings and monograph collections will generally be edited by a Guest editor.

High quality, novelty of the content and potential for the applications to modern problems in applied science will be the guidelines for the selection of the content of this series.

Instructions for Authors

Submission of proposals should be addressed to the editors-in-charge or to any member of the editorial board. In the latter, the authors should also notify the proposal to one of the editors-in-charge. Acceptance of books and lecture notes will generally be based on the description of the general content and scope of the book or lecture notes as well as on sample of the parts judged to be more significantly by the authors.

Acceptance of proceedings will be based on relevance of the topics and of the lecturers contributing to the volume.

Acceptance of monograph collections will be based on relevance of the subject and of the authors contributing to the volume.

Authors are urged, in order to avoid re-typing, not to begin the final preparation of the text until they received the publisher's guidelines. They will receive from World Scientific the instructions for preparing camera-ready manuscript.

www.ingramcontent.com/pod-product-compliance
Lightning Source LLC
Chambersburg PA
CBHW050636190326
41458CB00008B/2299

* 9 7 8 9 8 1 0 2 0 4 6 6 2 *